SpringerBriefs in Materials

More information about this series at http://www.springer.com/series/10111

Joe Briscoe • Steve Dunn

Nanostructured Piezoelectric Energy Harvesters

 Springer

Joe Briscoe
School of Engineering
and Materials Science
Queen Mary University of London
London, UK

Steve Dunn
School of Engineering
and Materials Science
Queen Mary University of London
London, UK

ISSN 2192-1091 ISSN 2192-1105 (electronic)
ISBN 978-3-319-09631-5 ISBN 978-3-319-09632-2 (eBook)
DOI 10.1007/978-3-319-09632-2
Springer Cham Heidelberg New York Dordrecht London

Library of Congress Control Number: 2014946580

Printed on acid-free paper

Springer is part of Springer Science+Business Media (www.springer.com)

Preface

Great interest and considerable research funding has focused on the development of engineered systems that convert stray energy into useful energy to increase energy security or mitigate further increases in atmospheric CO_2 levels. We are all familiar with photovoltaic cells that convert sunlight into electricity, solar to heat systems for heating and hot water, wind and sea turbines that convert a flowing fluid into energy and natural systems such as photosynthesis in plants that produce hydrocarbons from H_2O, CO_2 and sunlight.

However, this book instead focuses on devices that convert stray vibrations and movement directly into electricity. This process can take many forms and is the basis for most power generation around the world where steam is used to spin a turbine and thereby produce power. These devices rely on a variety of physical processes that can convert energy from one form to another, such as capacitive, electrostatic or electromagnetic systems, or the use of piezoelectric materials at a variety of length scales. Due to large number of systems available this book will focus on the new breed of nanostructured piezoelectric generators for small-scale device applications, and aims to provide the reader with a brief history and highlights some of the recent developments in understanding.

A piezoelectric material benefits from a crystallographic peculiarity in which a dipole is formed when the material is pressed. Without knowing it many of us have experienced this in our daily lives when lighting a gas cooker, or driving a diesel car, or perhaps when having an ultrasound scan. In these cases large or macroscopic crystals of the piezoelectric material are manufactured to give the unique response that is required in the device. When the crystal is miniaturised it remains piezoelectric and so even a nanostructured system responds in the same way as a bulk system. Many piezoelectric materials lend themselves to processing at the nanoscale and it is this combination of availability of processing and unique materials properties that has led to the increased interest in nanostructured materials for kinetic to electrical energy conversion.

In this book we will briefly describe some of the physical phenomena associated with piezo (and ferro) electric materials, illustrate what types of devices have been produced and highlight where the future of the technology may lie. We will also try and demystify some of the processes used to produce devices and indicate how advances in processing can lead to the development of devices that have a number of niche applications.

London, UK   Joe Briscoe
                                                                  Steve Dunn

Contents

Chapter 1
Introduction

Functional materials are materials that encompass some inherent functionality. In other words they have a predictable, and perhaps, manageable change to a given response. For example a material that is used as a photodetector is functional in the sense that it responds in a predictable way to a given optical stimulus. A piezoelectric material is also functional. A sample of given dimensions will produce a known surface charge upon application of a given mechanical stress. This is known as the direct piezoelectric effect. Conversely if an electric field is applied to the piezoelectric material it will change dimensions in a process known as the converse piezoelectric effect. For the purpose of converting vibration (or kinetic energy) into electricity the direct piezoelectric effect is used. Devices have been using this curiosity of the material since early reports in the 1990s and early 2000s [1, 2] and as such it is very much an emerging field.

Due to the limitations of energy conversion many piezoelectric systems produce power that is on the order of milliwatts. This level of output is too small for large-scale system applications. Still, the use of piezoelectric materials or devices based on piezoelectric materials to harvest power is becoming increasingly popular with piezoelectric elements being embedded in the ground to recover the energy of footsteps; Shenck and Paradiso at MIT embedded a system in shoes to scavenge heel strike energy [3]. A collaborative effort between the USA and Korea developed a micro-fabricated energy harvester using thin film PZT (a ferroelectric material) in 2005 [4].

Although the power outputs from devices are considered small, it is however sufficient to power small portable or hand-held electronic devices. In 2013 a British-based company, Perpetuum [5], announced large contracts to use vibration as a source of energy for structural health monitoring of railway rolling stock. This represented a significant breakthrough for such an emerging technology indicating that there are commercial applications available for whole-system solutions.

However, the systems mentioned above rely on macro or microscale devices. It is typical for those devices to use a variety of mechanical designs to maximise the mechanical stress. This is important as the amount of energy generated is directly

© Joe Briscoe and Steve Dunn 2014
J. Briscoe, S. Dunn, *Nanostructured Piezoelectric Energy Harvesters*,
SpringerBriefs in Materials, DOI 10.1007/978-3-319-09632-2_1

related to the stress applied to the system. Designs that have been investigated include a cantilevers and pressed disc-type systems. For nanostructured systems a different design approach has been adopted.

The work of Wang and Song [6] in 2006 is widely credited as the starting point for nanostructured piezoelectric harvesting systems. Since the discoveries of the mid-2000s significant progress has been made in developing a more detailed understanding of device performance characteristics. This increase in knowledge has led to increasing power output from devices and more robust test and measurement protocols.

This book aims to give the reader a basic introduction to the underlying physics and principles of piezo- and ferroelectric materials. These are materials that have been well known and described since the late 1800s and as such there is significant historical knowledge and understanding to be drawn upon. After this basic 'materials' introduction there will follow a more detailed review of the various approaches to nanostructured energy harvesting systems. In order to understand why nanostructured materials are of interest it is important to have a feel for the synthesis and manufacturing processes that are available for these systems. The majority of nanostructured piezoelectric devices use ZnO. ZnO is a particularly straightforward material to make as a nanostructure and there are well-known chemical processes that can make nanorods, nanowires, nanoflowers and even nanoshrimps of ZnO. However, more recently both PZT and $BaTiO_3$—well-known ferro- and piezoelectric materials—have been shown to be suitable for energy harvesting at the nanoscale. Such work has also shown that using a material with higher energy conversion parameters leads to enhanced energy generation, as predicted from fundamental theory.

References

1. Henty DL (1998) US Patent 5,838,138
2. White NM, Glynne-Jones P, Beeby SP (2001) A novel thick-film piezoelectric micro-generator. Smart Mater Struct 10:850–852. doi:10.1088/0964-1726/10/4/403
3. Shenck NS, Paradiso JA (2001) Energy scavenging with shoe-mounted piezoelectrics. Micro, IEEE 21:30–42. doi:10.1109/40.928763
4. Jeon YB, Sood R, Jeong J-h, Kim S-G (2005) MEMS power generator with transverse mode thin film PZT. Sens Actuators A Phys 122:16–22. doi:10.1016/j.sna.2004.12.032
5. Perpetuum. http://www.perpetuum.com/
6. Wang ZL, Song J (2006) Piezoelectric nanogenerators based on zinc oxide nanowire arrays. Science 312:242–246. doi:10.1126/science.1124005

Chapter 2
Piezoelectricity and Ferroelectricity

2.1 Background

Piezoelectricity is electric charge that accumulates in response to applied mechanical stress in materials that have non-centrosymmetric crystal structures. Piezoelectricity was discovered in the late 1800s by French physicists Jacques and Pierre Curie [1]. A subset of piezoelectricity is ferroelectricity: see Fig. 2.1. As a result of this all ferroelectric materials are piezoelectric. Ferroelectric materials exhibit interesting semiconductor properties that are analogous to the properties found in stressed piezoelectric materials. There is, therefore, a need to have a basic understanding of piezo- and ferroelectric materials as well as the relationship between them when developing a piezoelectric-based energy harvesting system.

The piezoelectric effect can be described as a linear interaction between the mechanical and the electrical state of a material that has no inversion symmetry within the crystal. It is in effect an electromechanical process. Materials lacking a centre of inversion are often termed non-centrosymmetric. Of the 32 crystal classes 20 possess direct piezoelectricity, and 10 of these 20 are polar crystals, exhibiting spontaneous polarisation in the absence of mechanical stress as the dipole moment associated with the unit cell has a non-vanishing component. These crystal structures will exhibit pyroelectricity—the generation of a charge upon the presence of an oscillating thermal gradient. Additionally if the dipole moment is reversible under the application of a sufficiently large electric field, the material is ferroelectric. The piezoelectric effect converts mechanical and electrical energy in both directions for any material: if a material exhibits a direct piezoelectric effect, it will demonstrate converse piezoelectricity. The generation of charge upon application of a stress is the direct effect while the converse effect results in a mechanical strain developing from the application of an electrical field. For example, a ferroelectric material $PbZr_xTi_{1-x}O_3$ (PZT) will produce a measurable piezoelectric output at 0.1 % deformation.

© Joe Briscoe and Steve Dunn 2014
J. Briscoe, S. Dunn, *Nanostructured Piezoelectric Energy Harvesters*,
SpringerBriefs in Materials, DOI 10.1007/978-3-319-09632-2_2

3

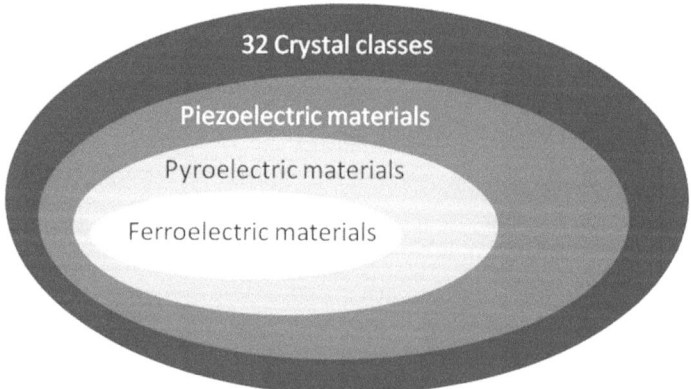

Fig. 2.1 Relationship of piezo, pyro and ferroelectric materials

The extent of the charge developed is limited by the mechanical coupling efficiency of the material: an often much debated constant that relates the ratio of mechanical energy to electrical energy conversion in the material.

2.2 Polarisation

When considering a piezoelectric material it is important to consider the change of polarisation P under mechanical stress. There are a number of possible causes for the development, or structuring, of the dipole due to the external stress. An additional complication is that piezoelectricity in the crystal is then able to be produced at different magnitudes and directions that are dependent on criteria that are both material and stress determined. The direction and strength of the polarisation are dependent on three factors. The first is the orientation of P within the crystal, the second is the crystal symmetry, and third is the stress applied by mechanical deformation of the system. Any variation in P can be measured as the change in surface charge density at the crystal faces. This is the surface polarisation with units of Cm^{-2} more commonly seen in literature as $\mu C/cm^2$. This difference in the electrical field between faces of a sample is caused by a change in dipole density. A commonly cited example is a cube of quartz with an applied load of 2 kN producing 12,500 V of potential difference [1]. This is the source of many 'sparking' systems, and some quartz based clocks which use the resonance of vibration as a measure of time.

Piezoelectricity results from a combination of the electrical behaviour of the material with Hooke's law as shown below.

The electrical behaviour of a material can be described by:

$$D = \varepsilon E, \tag{2.1}$$

where the displacement (D) of charge density, ε is permittivity and the electric field strength (E) that is applied to the sample.

Hooke's Law states:

$$S = sT, \tag{2.2}$$

where the strain (S), the compliance (s) and the stress (T) are used to define the system.

Equations (2.1) and (2.2) can be combined to form new relationships which are in the strain charge form:

$$\{S\} = \left[s^E \right] \{T\} + \left[d \right] \{E\} \tag{2.3}$$

$$\{D\} = \left[d^t \right] \{T\} + \left[\varepsilon^T \right] \{E\}, \tag{2.4}$$

where $[d]$ is the direct piezoelectric effect matrix and $[d^t]$ is the matrix used to define the converse piezoelectric effect: E indicates that a zero, or constant, electric field is found in the system and T indicates a zero, or constant, stress field across the system. The transposition matrix is determined by t.

There are a total of four piezoelectric coefficients, d_{ij}, e_{ij}, g_{ij}, h_{ij}, which are defined by convention as follows:

$$d_{ij} = \left(\frac{\partial D_i}{\partial T_j} \right)^E = \left(\frac{\partial S_j}{\partial E_i} \right)^T \tag{2.5}$$

$$e_{ij} = \left(\frac{\partial D_i}{\partial S_j} \right)^E = \left(\frac{\partial T_j}{\partial E_i} \right)^S \tag{2.6}$$

$$g_{ij} = \left(\frac{\partial E_i}{\partial T_j} \right)^D = \left(\frac{\partial S_j}{\partial D_i} \right)^T \tag{2.7}$$

$$h_{ij} = \left(\frac{\partial E_i}{\partial S_j} \right)^D = \left(\frac{\partial T_j}{\partial D_i} \right)^S \tag{2.8}$$

In Eqs. (2.5)–(2.8) the first terms are related to the direct piezoelectric effect and the second correspond to the converse piezoelectric effect. In a piezoelectric material, where the polarisation is crystal-field induced, there is a convention that makes it possible to calculate piezoelectric constants d_{ij} from an electrostatic lattice, which is formally determined by the crystal lattice. Alternatively, it has been shown that this can be achieved by using high order Madelung constants [2, 3].

Fig. 2.2 Polarisation of dielectric material under an external electric field

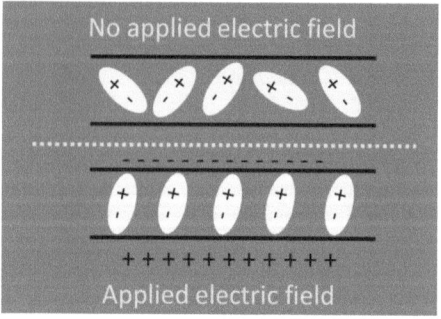

As a general rule when materials are polarised, P is proportional to any applied external field E, in this case an electric field. So polarisation is considered to be a linear function, until mechanical breakdown of the sample or some other limiting factor. While there are a number of terms to be found in the recent literature for the polarisation developed in a material under an applied electric field convention dictates that this is termed 'dielectric polarisation' (see Fig. 2.2). A group of materials termed paraelectric show nonlinear polarisation. This behaviour arises from the nonlinear permittivity of the materials. Therefore, the corresponding slope of the polarisation curve is not constant as is found for standard dielectric materials but is a function of the external electric field. This relationship means that as the sample develops P through the applied load there is a change in the permittivity and so nonlinear behaviour is formed.

2.3   Ferroelectricity

As mentioned earlier there are a number of crystal classes that exhibit piezoelectricity and within that group a subset are ferroelectric. Ferroelectric materials are of interest here for a number of reasons. The first is that ferroelectric materials tend to exhibit some of the largest electromechanical coupling coefficients and so could produce more effective energy harvesting devices, and secondly there is a great deal known about the semiconducting properties of ferroelectric materials. This is of interest as when stressed there are periods when a piezoelectric material will behave in an analogous way to a ferroelectric material due to similarities of dipole formation.

In addition to exhibiting nonlinear polarisation, ferroelectric materials will spontaneously demonstrate polarisation at zero applied electric field (see Fig. 2.3). The key and distinguishing feature for a ferroelectric material is the reversibility of the polarisation through the application of an applied external electric field. The term ferroelectric comes from a historical representation of magnetic hysteresis loops in ferromagnetic materials, hence the term ferroelectric. A ferroelectric hysteresis loop is dependent on a number of factors such as the number of cycles the sample has

Fig. 2.3 Spontaneous
polarisation for a poled
ferroelectric material

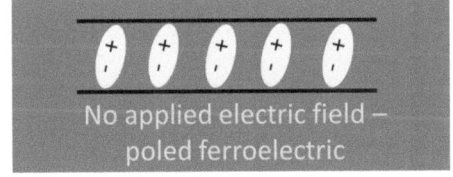

been subjected to. A large number of electric field cycles can reduce the response of the material through fatigue and so the polarisation can be history dependent.

A ferroelectric material will only demonstrate ferroelectric behaviour below a critical phase transition temperature, termed the Curie temperature, T_c. Above this temperature the materials are paraelectric and behave in a nonlinear manner. The value of T_c is material dependent and when designing an energy harvesting the operating temperature would be limited by T_c as above that temperature stress will not result in a measureable P.

When a material has two or more polarisation states that can exist in the absence of an applied electric field and when these stable polarisations can be changed to another stable state by the application of an external electric field it is said to be ferroelectric. This feature was first discovered by Valasek in 1920 for the material Rochelle's salt. Through detailed investigations Valasek recognised that the crystals had a similar nature and behaviour to ferromagnetic materials such as iron. Rochelle's salt displayed a Curie temperature above which the unusual properties disappeared and demonstrated large dielectric and piezoelectric response below the Curie temperature. As mentioned earlier the use of the term ferroelectric describes materials that exhibit ferroelectric behaviour as they were believed to be the electrical equivalent of a ferromagnetic material. It should be noted that the two phenomena have very distinct origins, but that many of conventions and terminology associated with ferroelectric materials owe their roots to ferromagnetism. In the case of ferromagnetism the property stems from electron spin and angular momentum of unpaired electrons that alter the magnetic moment of atoms within the crystal. The spontaneous polarisation in ferroelectric materials forms as a result of crystal asymmetry below the Curie point.

Another term that made the transition from ferromagnetic to ferroelectric materials is that of a domain, or region of aligned polarisation for a ferroelectric material. As such the spontaneous polarisation for a ferroelectric material is the magnitude of the polarisation in a single domain. In a typical system, due to thermodynamic constraints, the surface of a sample will have a random array of domains to reduce the surface free energy. The spontaneous polarisation of a ferroelectric material can range over 4 orders of magnitude from Rochelle's salt with a spontaneous polarisation of 2.5×10^{-2} µC cm^{-2} to materials like lithium niobate with spontaneous polarisation measured at 78 µC cm^{-2} [4].

2.3.1 Phase Transition

The phase transition that occurs at the Curie point responsible for formation of spontaneous polarisation in ferroelectric crystals is characterised as order–disorder or displacive. In displacive phase transitions atomic displacements in the paraelectric phase are oscillations around a non-polar site, while below the Curie point they are about a polar site. The classes of displacive ferroelectrics include a large range of materials such as the ABO_3-based perovskite structures which include $BaTiO_3$, $PbTiO_3$ and $PbZr_xTi_{1-x}O_3$ (PZT). A perovskite structure can exist in two crystallographic forms. Above the Curie point the crystal class is body centred cubic and the unit cell is centrosymmetric with the positive and negative charges coinciding and so being a purely paraelectric material. In the case of $BaTiO_3$ the unit cell comprises barium ions located at corner sites and a body centred titanium cation residing within a coordination polyhedron of oxygen anions (Fig. 2.4). When a sample is cooled through the Curie point (around 120 °C for $BaTiO_3$) a phase transition occurs changing the crystal class from body centred cubic to tetragonal. In the case of barium titanate this is a result of oxygen anions and barium cations being shifted in the same direction relative to the location of the titanium cation. This occurs without significant distortion of the oxygen octahedra. The phase transition moves the titanium cation away from the centre of the unit cell resulting in a dipole in the unit cell.

In the paraelectric phase order-disorder transitions for the atomic displacements are about a double-well or, in some cases a multi-well configuration of sites. In contrast below T_c or in the ferroelectric phase atomic displacements are ordered on a subset of the available wells for the paraelectric phase. This results in long range ordering taking place for a ferroelectric crystal at temperatures below T_c and leads to the creation of spontaneous polarisation and domain structures within the material.

2.3.2 Spontaneous Polarisation

The spontaneous polarisation for a ferroelectric material can develop in opposite directions along at least one axis giving rise to distinct sections of the sample that exhibit the same polarisation. Within a crystal such regions can coexist and will

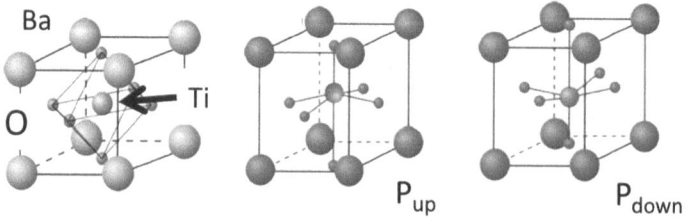

Fig. 2.4 Crystal structure for $BaTiO_3$ showing effect of Ti^{4+} displacement on spontaneous polarisation of the lattice

only differ in the orientation of the polarisation. These regions within the crystal are termed ferroelectric domains. Convention and matrix symmetry dictate that any domain that is orientated along the Z-axis will be described as a C domain. Additionally they are known as a C+ domain if they align to the Z+ direction and C– if they align along the Z– direction. A lateral domain is orientated along the plane across the surface in the direction of measurement. A domain wall is a region of local crystalline, and in some cases material, disorder that separates adjacent domains. The cross section of a domain wall has been reported to range from 0.5 nm to several nm. A 180° domain wall will separate C+ and C– domains while a domain wall that separates a lateral and a C domain is a 90° wall.

Spontaneous polarisation is not homogeneous throughout a sample of ferroelectric material and environment surrounding the sample. At the interface between the ferroelectric and surrounding environment the spontaneous polarisation rapidly reaches zero. Immediately outside the ferroelectric the polarisation decays to zero and at a non-zero value, determined by the material and its electrical history, termed the spontaneous polarisation inside the material. The net effect of this large change in the polarisation is to produce bound charge at the interface between the ferroelectric and the surroundings. A ferroelectric material's surface as with all surfaces is defected and these defects influence the local electron density in the materials structure. This leads to differences in polarisation around local defects. The divergence of spontaneous polarisation at the interface leads to a depolarisation field. In order to maintain charge neutrality there needs to be an electrical field that opposes the spontaneous polarisation. The depolarisation field significantly impacts the physical properties of a ferroelectric material. They can act to suppress and in some cases can fully destroy the ferroelectric state due to a build-up of surface charge.

In the case where there is the formation of a single domain in a material this creates a large depolarisation field. Such a large depolarisation field causes inherent instability in the sample. To compensate for this a native, or unmodified, sample of ferroelectric material will form in a structured that has multiple oppositely orientated domains. This process occurs spontaneously at temperatures below T_c and minimises the total free energy (TFE) of the system. In the case of a ferroelectric material we can define TFE as the total energies associated with any depolarisation fields, surfaces, the energy of domain walls, and any other strain or crystal energies. Any stray unscreened depolarisation energy found in a polydomain material uses the movement of mobile charge carriers inside domains or external to the ferroelectric to complete compensation. This is particularly important when considering nanostructured materials for energy generation.

2.3.3 Measurements

The impact of the polarisation of a ferroelectric during the application of an external electric field is measured by a hysteresis loop (Fig. 2.5). The application of a low AC electric field causes domains in the material to align with the applied field and

Fig. 2.5 Hysteresis loop for
a ferroelectric material,
where E_c is the coercive field
and P_r is the remnant
polarisation

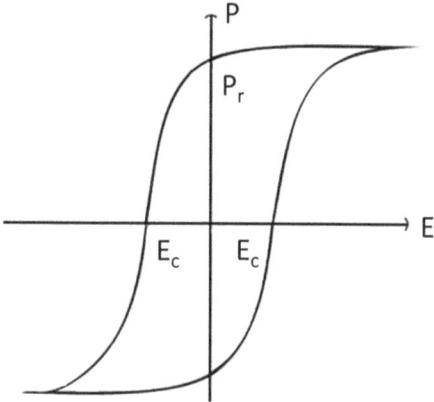

causes polarisation to increase linearly during the initial stages of the ramp in the electric field. As the electric field is increased, previously unfavourably orientated domains change polarisation and align along the electric field. This generates in an increase in the net polarisation of the system. When the electric field is released a number of domains that have orientated to the direction of the applied field remain. This remnant polarisation (P_r) is a feature of the material and sample geometry. In some exceptional cases the applied electric field is sufficient to fully align all domains in the sample. In these cases P_r will equal the spontaneous polarisation of the material and in this case the material has a saturation remnant polarisation. The reversal of the applied electric field applied to the sample at a sufficient level causes the polarisation in the crystal to be reversed. An intermediate step to the reversal of spontaneous polarisation is when the material exhibits zero polarisation. This electric field, required to reduce polarisation of the sample to zero, is termed the coercive field (E_c). When an applied electric field is increased above E_c this generates an opposite direction in the sample.

2.3.4 Size Influences on Ferroelectric Behaviour

It is accepted that ferroelectric materials demonstrate some influence of size on the ferroelectric properties. The origins of this are not fully understood but it has been suggested they come from a combination of factors. The size strongly influences the density of domain walls and the associated energy associated with domain wall formation. The decrease in the volume of a ferroelectric sample is associated with increasing energy associated with the domain wall formation and stabilisation. The reduction in size influences ferroelectrics that form polydomains to form a single domain below a threshold size. Barium titanate particles show a change from polydomain to single domain below around 100 nm. The formation of single domains significantly increases any depolarisation field that is associated with the sample of ferroelectric.

If the particles cannot compensate the increase in depolarisation energy through the migration of internal or external charge carriers, the ferroelectricity of the material will be destroyed.

A second influence thought to drive the size effects in ferroelectric materials are dipole interactions that are exhibited over a long range. These interactions support ferroelectric phase formation. Using Ginzburg–Landau theory it has been shown that there are size constraints below which ferroelectric domain structure collapses. The free energy of an inhomogeneous ferroelectric particle over a range of sizes where polarisation has been used as a key parameter in the modelling has been used to demonstrate this in some detail. Accordingly the polarisation is expressed by two key factors. The correlation length is the average coordination distance for polarisation fluctuations in the material. When a large correlation length exists the dipoles found in the crystal will interact over long distances within the lattice. The extrapolation length relates the difference in coupling strength found at the bulk and surface of the material. When coupling strength at the surface is lower than that found in the body of the sample the surface will be significantly disordered in comparison with bulk. The correlation length determines the distance of the disorder penetration into the material. As the range of disorder in the surface influences the bulk this is dictated by the correlation length of the material. The change in ratio of surface to bulk as a material's size is reduced means that eventually the bulk of the material is determined by the surface disorder. Further reduction in size of the sample means that the disorder in the system destroys the polar state.

2.3.5 Ferroelastic Behaviour

The definition of strain is deformation of particles within a body of material that is experiencing an applied stress when compared to the original position of the sample. Ferroelasticity is a property of a ferroelectric material where a spontaneous strain develops due to stress–strain behaviour below T_c. In an analogous statement to that of ferroelectric materials, a sample exhibits ferroelastic behaviour when there are two or more orientation states available in the absence of a mechanical stress. In essence the states in the sample can be changed by applying a sufficient level of mechanical stress. In a material such as an ABO_3 perovskite that exhibits ferroelastic behaviour the displacement of atoms in the crystal lattice produces a dipole related to the spontaneous polarisation and strain in the crystal. When there is no applied electric field the mechanical stress of the system can produce ferroelastic switching in the material.

In an unpoled ferroelectric material domains form in all available axes in order to minimise the energy associated with the polarisation of each domain. The application of mechanical stress to a system causes equal switching in all axes of the sample which produces an increase in strain but no associated change in polarisation. After a sample of material has been subjected to an electric field at a level required for polarisation of the sample it can produce a single domain or specific domain structure. The subsequent application of mechanical stress can produce ferroelastic domain switching in the sample. Such a process can cause the sample to become depolarised or form a domain structure that consists of a large number of domains. Such a process of depolarisation associated with stress has been demonstrated in thin film samples of a variety of perovskite type materials including PZT.

2.4   Semiconductor Materials

Molecular orbital theory describes the allowed and forbidden energy states for electrons in a solid material. A large range of materials properties are determined by the electronic band structure and include electrical conduction, optical properties, magnetic properties and heat conduction. For a single atom electrons occupy atomic orbitals that are associated with discrete energy levels. Every energy level is able to hold at most two electrons that, according to the Pauli Exclusion Principle, must have opposite spin. When more atoms are chemically bonded molecular orbitals form which are a combination of individual atomic orbitals. These molecular orbitals consist of bonding and non-bonding orbitals. It is the arrangement of these orbitals that determines many of the fundamental properties of a material.

Constructive interference of orbitals produces a lower energy state or bonding orbital, destructive interference produces an orbital (anti-bonding) that has higher energy than the atomic orbital from which it has been formed. The number of molecular orbitals produced shows proportionality to the number of atoms that make up the solid, so a large number of molecular orbitals will be formed when there are collections of large numbers of atoms. As bonding orbitals increase the energy levels that are formed converge, eventually becoming indistinguishable and forming continuous energy bands. In a semiconductor when a discrete band gap is formed the bonding orbitals form a band known as the valence band (E_v) while anti-bonding orbitals form a band known as the conduction band (E_c). In a metallic system there is no distinguishable band gap and the top of the bonding orbitals is known as the Fermi level (E_f) with an overlap into anti-bonding orbitals leading to metallic properties. The differences between semiconductors and insulators are often difficult to discern but a band gap exceeding 3.5 or 4 eV leads to primarily insulator behaviour as terrestrial sunlight is unlikely to excite a band transition.

We can use some broad definitions to enable us to understand the interplay between a metal and insulator material. If a current is produced when a moderate electric field is applied to a material, it is termed a conductor. Conduction is a process that stems from the partially filled conduction band, or overlaps with unfilled anti-bonding orbitals enabling mobile electrons to move through the lattice. A metal such as lithium has a partially filled conduction band due to the overlap between the bottom of the anti-bonding orbitals and top of the bonding orbitals. A material upon application of an external electric field that does not produce a current is called an insulator. For an insulating material the forbidden energy states produce a gap that is found between the E_v and E_c which is called the band gap (E_g). In the case of an insulator electrons can only occupy E_c if they are excited over the band gap from E_v. This requires energy to be absorbed that is at least equal to or exceeding E_g. An insulator will typically have a wide band gap such as for diamond (E_g ca. 7 eV) which means that under ambient conditions it is not possible for excitation of electrons to occur. If a material has E_g below 2 eV such as Si (E_g ca. 1 eV) electrons are easily promoted from E_v to E_c through the adsorption of photons or thermal excitation producing mobile electrons and so conductor type behaviour. It would be typical to consider such materials as a semiconductor.

2.4.1 Intrinsic and Extrinsic Properties

If the structure of a semiconductor at a crystalline level contains few impurities, defects or in exceptional cases is impurity free, then the semiconductor is intrinsic. The number and density of charge carriers as well as the majority carrier type produced for the material is determined purely by the inherent materials properties. It is also possible for impurities to be added to the material in a process known as doping. A semiconductor that has its properties altered in this way is described as extrinsic. A dopant is any element other than one that would usually be found in the semiconductor structure. These impurities create a defect by replacing an atom in the crystal structure. The doping of a semiconductor alters many of the fundamental properties of the material by influencing the concentration of charge carriers—either electrons or holes—and hence can change the majority carrier type, concentration or mobility.

When the dopant atom has more valence electrons than that of the atoms typical of the crystal structure extra electron density can be donated into the E_c. Dopants of this nature are termed donors as there is an increase in the concentration of electrons in the E_c and the material is said to be an n-type semiconductor. When the system has been donor-doped E_f moves towards the conduction band. When a dopant has fewer valence electrons they are known as acceptor dopants. In this case the dopant atom will remove electron density from the E_v which leads to an increase in the number of holes available for conduction. A dopant with this nature produces a p-type material and E_f moves down towards E_v.

Early contributions by Shockley indicated that when the crystal structure of a semiconductor is disrupted a series of new energy levels can be created. These are known as surface states and have effectively been removed from E_c so there has not been the creation of new energy states in the semiconductor. Surface states are of interest as they remove or trap carriers from the body of the sample until the situation where the energy associated with E_f at the interface and the energy in the body of the material have reached equilibrium. By determining where the surface energy states lie relative to E_f this determines whether the surface state can effectively trap electron or holes. Any surface state with energy greater than E_f will trap holes and surface states below E_f will trap electrons. The trapping of a mobile species at the surface produces bound charge, in this case electrostatic, such that after formation of the surface states the sample will be electrically neutral.

Explaining this further we will consider the action of depolarisation fields that are found inside a semiconductor. These fields compensate the local electric field by drawing any free carriers that are oppositely charged towards the surface. This results in increased carrier density in regions at the interface which are known as the space charge regions (SCR). A positive surface potential will draw electron density towards the interface and the SCR is termed an accumulation region. Associated with an increase in electron density is a conceptual downward band bending of the semiconductor. At a negative surface the local potential draws holes towards the surface and a depletion region is formed with an associated upward bending of the band structure.

A definition of the work function for a semiconductor is the smallest energy required to remove an electron to a point immediately outside the sample. In cases where it is useful to determine the work function it can be calculated as the difference between E_f and the vacuum level known as E_o. In cases where the work function for a given metal (ϕ_m) and a semiconductor (ϕ_s) are not the same when they are brought into contact electrons flow over the contact. They will flow from the lowest E_f until equilibrium is reached. Such a system is known as a Schottky barrier, with the height (ϕ_b) being the barrier that electrons have to overcome to flow between materials (see Fig. 2.6).

In cases where ϕ_m is smaller than ϕ_s and ϕ_b is effectively negative, then an ohmic contact will be formed as an electron will flow across the interface from the metallic side to the semiconductor. Alternatively in cases where the Schottky barrier height is positive a rectifying contact will be formed with an electron being transferred from the semiconductor to the metallic side. This transfer of electrons across the interface causes a depletion region to form with the associated upward bending for the semiconductor.

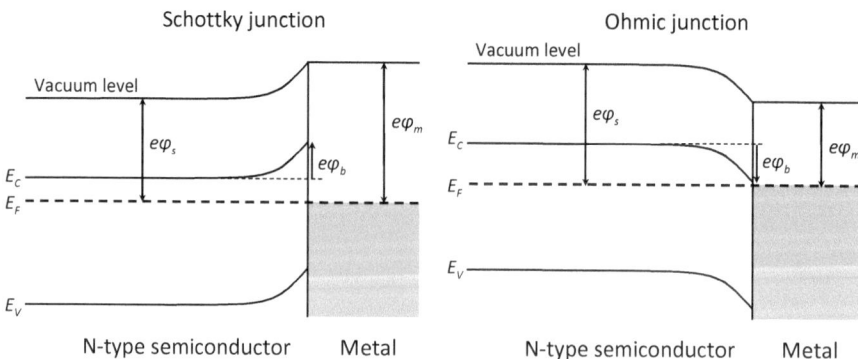

Fig. 2.6 Band diagrams of Schottky and Ohmic junctions between a metal and an n-type semiconductor. For a Schottky junction the metal work function (ϕ_m) is larger than that of the semiconductor (ϕ_s) leading to upward band bending of the semiconductor and formation of a depletion region at the interface. There is a positive energy barrier ($e\phi_b$) at the interface. For an Ohmic junction ϕ_m is smaller than ϕ_s leading to downward band bending in the semiconductor, and a negative energy barrier at the interface

2.5 Ferroelectric and Piezoelectric Materials as Semiconductors

There is an historical concept of treating a ferroelectric material as an insulator. There is now a growing understanding that although carrier concentrations and therefore conductivity may be low for many ferroelectric materials, a ferroelectric can be considered to be a wide band gap semiconductor. This has been demonstrated in perovskite and non-perovskite materials and in these cases it has been shown that the spontaneous polarisation will modify and to a large extent determine the electronic properties. For example, a typical lead-based ferroelectric such as PZT will form a rectifying Schottky contact with platinum. This is a property that is associated with typical semiconductor–metal junctions. Further evidence of the semiconductor properties of a PZT sample have been demonstrated by current–voltage and capacitance–voltage measurement techniques. Upon the application of a small DC bias across a ferroelectric sample in contact with metal electrodes it has been shown that semiconductor properties of the sample can dominate ferroelectric properties. Calculations show that the space charge region grew and with this there was a significant reduction in the ferroelectric nature of the sample [5]. It is now well accepted that under super band gap illumination ferroelectric materials follow semiconductor theory. They produce a photo-induced state away from equilibrium that results from the formation of photogenerated electron-holes. There is a movement of photoexcited charge carriers within the ferroelectric that screen any surface charge. This has the additional effect of producing band bending analogous to that found and associated with the surface states of a non-polar semiconductor material.

The electrostatic interactions for a ferroelectric interface are described by the formation of charge in response to P and the associated screening of that charge by the depolarisation field $-\rho$ within the sample. There are four possible situations to determine the influence of screening on depolarisation [6].

1. Surface charge is completely unscreened, $P = -\rho$
2. Surface is partially screened, $P > -\rho$
3. Surface charge is completely screened, $-\rho = 0$
4. Surface charge is over screened $P < -\rho$

The depolarisation field can effectively screen any surface potential by determining the movement of the mobile species—electrons or holes—from within the bulk of the sample (Fig. 2.7). In the case of external screening this can occur when ionic species or polar molecules interact with the ferroelectric surface and become adsorbed into the electric Stern layer. This structure is formed at the interface of a solid sample when placed in a liquid. It can also exist if there is a sufficiently thick layer of water or other suitable solvent on the ferroelectric. Internal screening occurs when an electron or hole is driven to the interface of the ferroelectric and the surroundings.

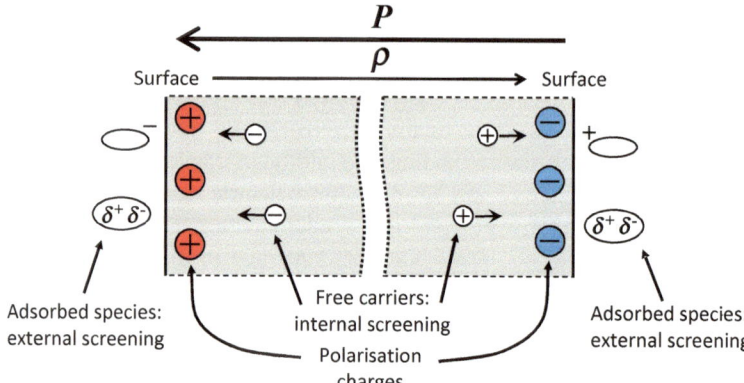

Fig. 2.7 Schematic of screening in a ferroelectric material. Both free carriers internal to the ferroelectric material and charged species in the environment external to the material move in response to the depolarisation field, ρ, that arises due to the ferroelectric polarisation P. These mobile charges screen the surface potential arising due to the immobile polarisation charges

The case of totally unscreened surface charge is energetically unfavourable; a ferroelectric will only demonstrate this behaviour in exceptional circumstances such as when being held under high vacuum as in this case there are few species available to provide external screening of the surface charge. A sample can become over screened, such as the case when the interaction with charged species is over-compensating the surface potential, which is not typical for ambient conditions. This is generally only observed under specific circumstances such as when an electric field is applied. More typically the ferroelectric surface is only partially or fully screened. The interaction of the extent of external and internal screening is dependent upon a number of factors. These include features such as the non-equilibrium carriers produced by interaction with photons that are greater in energy than the band gap, the thermally excited states or defects associated with the surface of the system. In a material with a large number of defects, such as PZT, surface charge is predominantly screened by mobile electrons or holes from within the lattice. The opposite can be found for a material that has a low number of defects such as lithium niobate. In this case the lack of mobile carriers means that the predominant screening process is the absorption of charged species or ions.

References

1. Jaffe B, Cook JM, Jaffe H (1971) Piezoelectric ceramics. Academic, London
2. Kittel C (2004) Introduction to solid state physics, 8th edn. Wiley, New York
3. Scrymgeour DA, Hsu JWP (2008) Correlated piezoelectric and electrical properties in individual ZnO nanorods. Nano Lett 8:2204–2209. doi:10.1021/nl080704n

4. Stock M, Dunn S (2011) LiNbO3 - a new material for artificial photosynthesis. Ultrason Ferroelectr Freq Control IEEE Trans 58:1988–1993. doi:10.1109/TUFFC.2011.2042
5. Tiwari D, Dunn S (2009) Photochemistry on a polarisable semi-conductor: what do we understand today? J Mater Sci 44:5063–5079
6. Dunn S, Tiwari D, Jones PM, Gallardo DE (2007) Insights into the relationship between inherent materials properties of PZT and photochemistry for the development of nanostructured silver. J Mater Chem 17:4460–4463

Chapter 3
Nanostructured Materials

As discussed in the previous chapter, the study of piezoelectric materials dates back over a century, and a huge number of different materials displaying piezoelectric behaviour have been demonstrated. However, to date there has been relatively limited use of nanostructured piezoelectric materials in functioning devices. This is partly because understanding of the nanoscale size effects on ferro- and piezo-electricity is still being developed, as discussed in Sect. 2.3.4. It is also because production of nano-sized structures of piezoelectric materials often involves complex processing either due to the ternary, quaternary or higher number of elements combined in specific quantities, or the ceramic nature of a number of piezoelectric materials, requiring high processing temperatures. However, there are some exceptions to this, the most well known being zinc oxide (ZnO). As discussed below, a number of simple methods have been extensively studied for the production of nanoscale ZnO, and as such it is the most widely used material in piezoelectric nanogenerators. However, other well-known materials have also been investigated including lead zirconate titanate (PZT) and barium titanate. In this chapter the synthesis methods used to produce these materials are summarised, and details of the development and specific examples of piezoelectric nanogenerators are described.

3.1 Synthesis

Zinc oxide readily forms a range of nanostructures, especially 1D-type structures such as rods and wires due to a preferential growth along the c-axis of its wurtzite structure (see Fig. 3.1). As such there are a variety of methods for synthesising nanostructured ZnO, and it has been widely used in nanogenerators. The most common synthesis methods are detailed below along with other examples of nanostructured piezoelectric materials that have been produced.

© Joe Briscoe and Steve Dunn 2014   19
J. Briscoe, S. Dunn, *Nanostructured Piezoelectric Energy Harvesters*,
SpringerBriefs in Materials, DOI 10.1007/978-3-319-09632-2_3

Fig. 3.1 Crystal structure of
zinc oxide showing the
tetragonal arrangement of
zinc cations (*black*) and
oxygen anions (*white*)
forming a hexagonal wurtzite
unit cell

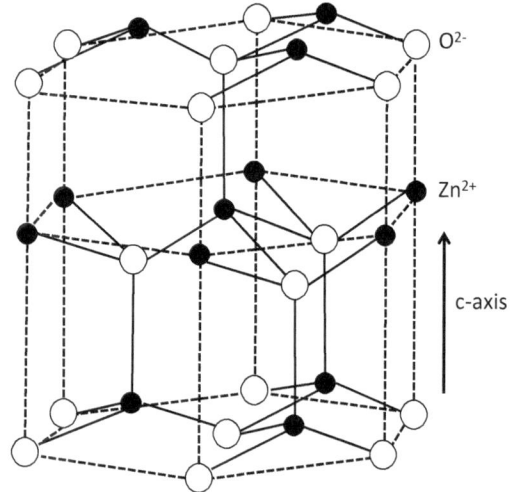

3.1.1   ZnO Nanostructures

A wide range of synthesis methods have been used to produce ZnO nanostructures
for more than a decade, and a number of review articles have been published on the
subject [1–3]. The main categories of growth method that are used are chemical
vapour, electrochemical and chemical bath deposition (sometimes referred to as
hydrothermal growth, though by convention this should also involve the use of high
pressure). The former method requires high temperatures (>400 °C), where the latter
two methods use aqueous chemical solutions and are performed at below 100 °C.
The key methodologies and parameters of these techniques are summarised below.

3.1.1.1   Chemical Vapour Deposition

To grow ZnO nanostructures by chemical vapour deposition (CVD), vapour species
are generated by the evaporation and/or chemical reduction of precursors in a reac-
tor such as a tube furnace. A gas flow transports the gaseous species to a solid substrate, which is at a lower temperature than the precursors. They nucleate on the
substrate generally either by the vapour–liquid–solid (VLS) or vapour–solid (VS)
mechanism. The VLS mechanism requires the substrate surface to be coated with a
metal catalyst, which is commonly gold in the case of ZnO growth [2]. At the elevated temperatures liquid gold droplets form in which the ZnO vapour dissolves.
When the droplets become supersaturated, the ZnO nucleates and begins to grow
outwards [2]. The nucleation of ZnO is controlled to form nanostructures of the
desired dimensions by controlling a number of parameters such as the initial metal
particle size [1], oxygen partial pressure and overall gas pressure [2].

VS mechanisms do not use a metal catalyst, but instead the substrate is coated with a ZnO thin film as a seed layer. For ZnO nanorods and wires, the carbothermal reduction of precursors is often used. In this method a 1:1 mix of commercial ZnO powder and graphite powder is heated between 700 and 900 °C, which reduces the ZnO and produces Zn vapour. A gas flow transports the vapour to a substrate, which is placed downstream at a slightly lower temperature point. By controlling the temperature, gas flow and position of the substrate, ZnO nanowires can be formed on the substrate with a good density and alignment [2, 4–6]. Alternatively, a precursor of Zn powder without graphite can be used [7] to directly produce Zn vapour by heating at 700–750 °C, and using a gas flow to again transport this to the substrate.

Although these vapour-phase methods can produce very well-aligned, highly crystalline ZnO nanorods or wires, the high temperatures required for the synthesis limit the substrates used to crystalline materials such as silicon, quartz or sapphire. As discussed below in Sect. 3.2.3, it is desirable to produce ZnO nanostructures on a range of substrates including flexible polymer materials, which are not suitable for such high-temperature synthesis methods. As such, low-temperature, solution-based methods are more commonly used.

3.1.1.2 Electrochemical Deposition

ZnO nanorods can be grown on almost any conductive substrate by electrochemical deposition. The conductive substrate is placed into a chemical bath, commonly containing zinc chloride and potassium chloride, with a counter electrode (Pt) and reference electrode (e.g. saturated calomel). The bath is heated to around 80–90 °C and a bias of around −1 V versus the reference electrode is applied to the substrate. This process has been used to successfully grow ZnO nanorods, generally for use in photovoltaic devices onto transparent conductive substrates [8–10], but is also suitable for growth of nanorods for nanogenerators.

3.1.1.3 Chemical Bath Deposition

Chemical bath deposition (CBD) is by far the most common method used for growth of ZnO nanorods for use in nanogenerators. CBD involves heating a mixture of an aqueous chemical zinc precursor with additives to alter the pH or control the formation of ZnO to produce the desired nanostructures. Most commonly a substrate seeded with a layer of ZnO is placed in an equimolar (0.01–0.1 M) mixture of zinc nitrate and hexamethylenetetramine (HMT), which is heated to 90 °C for 2–24 h to grow aligned ZnO nanorods on the substrate. This is based on the method developed by Vayssieres et al. [11], who demonstrated that this mixture (which had previously been shown to produce ZnO nanostructures [12]) could produce aligned arrays of rods nucleated on a substrate with aspect ratios around 10–20. In some cases the synthesis is repeated with a fresh chemical mixture to produce longer nanorods, as the reactants are depleted after some length of time due to homogeneous (in the solution) and heterogeneous (on the surface) nucleation of ZnO. In some cases additives such as poly(ethylenimene) (PEI) are used to produce elongated rods [13, 14], with aspect ratios of up to 200, or ammonium hydroxide [15] or ammonium chloride [16, 17] to produce rods more rapidly. Many reaction schemes have been proposed for the formation of ZnO using these methods, but it is generally assumed to proceed through a controlled hydrolysis of the Zn precursor, where HMT provides a controlled supply of hydroxide ions to maintain a [Zn^{2+}] to [OH^-] ratio that will lead to nanorod formation [18].

Alternatively, ZnO nanorods have also been produced from an aqueous solution of zinc nitrate by increasing the pH, most commonly using sodium hydroxide [19], which was studied in detail by Tak et al. [20]. This method produces ZnO nanorods on seeded substrates relatively rapidly, using a solution pH of 10–10.5, with high aspect ratios achievable by repeating the synthesis whereby the length of the nanorods increased more rapidly than the diameter, unlike the zinc nitrate—HMT method where the aspect ratio changes little with repeated syntheses.

It has been mentioned that for the chemical growth of ZnO nanorods, and in some cases for the vapour-phase growth, the substrate on which the nanorods are to grow must be pre-seeded with a layer of zinc oxide. This seeding layer is not commonly considered in detail, and for many nanogenerators the substrates are pre-seeded using a sputtering method such as rf-magnetron sputtering. However, it is also possible to seed substrates simply using chemical zinc precursors, commonly dissolved in alcoholic solvents. One such method used on transparent conducting oxides that was developed for photovoltaic applications is the repeated drop-casting, rinsing and drying of a 5 mM zinc acetate solution in ethanol followed by annealing at 350 °C [21]. This proceeds via the thermal decomposition of the zinc acetate on the substrate producing a layer of ZnO. However, as this method is not compatible with most polymeric substrates, similar alternatives have been developed such as the spin-coating of zinc acetate solutions onto polymer substrates followed by drying at only 100 °C [22–24]. Although this will not lead to the decomposition of the zinc acetate prior to ZnO nanorod synthesis, the zinc acetate layer is likely hydrolysed in the chemical bath and allows subsequent nucleation of ZnO on the surface. Seeding of cloth-based substrates is simplified as a precursor solution will easily soak into the cloth, so that the fibres can be well coated after only a short soak in a zinc acetate solution in ethanol or methanol [25].

Aqueous chemical solution-based methods therefore appear the most suitable for growth of ZnO nanostructures for nanogenerator applications, especially when combined with chemical seeding methods. These devices will not generate large quantities of power and are therefore best suited to off-grid applications such as remote sensors or portable charging. As such materials and synthesis cost must be kept to a minimum, which is best achieved by using simple, low-cost and scalable methods for synthesis of the active materials.

3.1.2 Lead Zirconate Titanate and Barium Titanate

With higher piezoelectric coefficients (see Sect. 2.2) than ZnO, lead zirconate titanate (PZT) and barium titanate (BTO) should be promising for use in piezoelectric nanogenerators. However, it has proved much more difficult to synthesise nanostructures of these materials, and as such reports of their use in nanogenerators are limited. One method for producing nanostructured PZT that has been applied to nanogenerators is electrospinning. In this process, most commonly applied to producing micro- or nanometre diameter polymer fibres, a precursor liquid is charged using a high voltage and then extruded through a small diameter needle forming a jet which collects on a grounded plate below the needle, forming the fine wires. This process was used to produce PZT fibres with an average diameter of 150 nm [26], which were later used in a nanogenerator [27]. The initial fibres were formed by mixing a PZT sol with poly(vinyl pyrrolidone) (PVP) to allow electrospinning. The electrospun PZT/PVP composite fibres were annealed at 650 °C to remove the PVP and crystallise the PZT in the perovskite structure. Due to their extremely high aspect ratio these fibres were deposited laterally on a substrate for use in a nanogenerator, rather than used as an aligned film, as described in Sect. 3.2.4. A similar electrospinning process using PZT/PVP precursors has also been used to fabricate nanogenerators using either lateral fibres [28], or rotated sections of fibres that form a film of PZT fibres embedded in polydimethylsiloxane (PDMS) aligned perpendicularly to the substrate [29]. These devices are discussed further in Sect. 3.2.4.

PZT and BTO nanorods have also been produced as aligned arrays on substrates similarly to ZnO nanostructures using a hydrothermal process. This involves the mixing of precursor chemicals in a solution which is heated above the boiling point of the solvent in a pressure-sealed container so that the final products are produced as a result of the reaction at elevated temperature and pressure. Such methods have been used to produce an array of PZT nanorods on a TiO_2-coated substrate [30], or to convert pre-formed TiO_2 nanorods into $BaTiO_3$. The relatively low-temperature synthesis (160–250 °C) of these structures suggests they could be used in a number of devices with similar configurations to ZnO, and a device using the $BaTiO_3$ nanorods has been demonstrated [31]. The hydrothermal method has been used to produce PZT nanorod arrays on niobium- or iron-doped strontium titanate (Nb:$SrTiO_3$, Fe:$SrTiO_3$) substrates, which were subsequently made into nanogenerators [32]. These substrates are both conductive and have a good epitaxial match to PZT, which led to well-aligned nanorod arrays, which helped with the fabrication of devices.

3.2 Device Architectures

The potential for energy harvesting from piezoelectric nanostructures was first demonstrated in 2006 by Wang and Song [33]. They found that as a conducting atomic-force microscope (AFM) tip was scanned across an array of ZnO nanorods grown

Fig. 3.2 Scanning-electron
microscope (SEM) micrograph
of ZnO nanorods grown using
the VLS method on sapphire
substrates from [33]

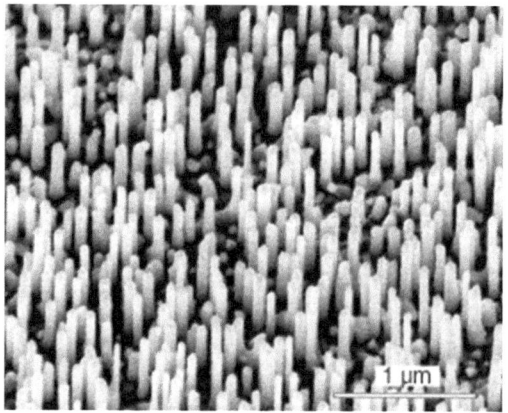

on a sapphire substrate a voltage output was produced each time the tip contacted
and then released a nanorod. They assigned this voltage signal to the piezoelectric
polarisation induced in the nanorod as a result of the strain caused by the tip.
Although the source of the output voltage in this system was initially controversial
[34, 35], later demonstrations of voltage and/or current output from arrays of ZnO
nanowires [36] led to significant growth in the study of nanostructured piezoelectric
materials for use in energy harvesting devices. In this section the evolution of these
energy harvesters is described including the key technological innovations and
progress in understanding of the working principles of the devices.

3.2.1   Early Investigation: Single Nanowire Testing

Early measurements of a piezoelectric voltage output from a strained nanorod were
gathered from measurements of single nanorods using a conductive AFM. In the
first study by Wang and Song [33] the ZnO nanorods were grown on a sapphire
substrate using a vapour–liquid solid (VLS) method (see Sect. 3.1.1). As well as
giving a very well-aligned, well-spaced array of nanorods due to the epitaxially
matched substrate and high temperature growth (see Fig. 3.2), this also resulted in
gold particles remaining on the tip of each rod after catalysing their growth. This
was extremely useful for the electrical measurements, as it allowed a good contact
between the conductive AFM tip and the top of the nanorod. In addition, Wang and
Song postulated that the Schottky barrier formed between the gold and ZnO was
important in producing a measurable voltage output of around 10 mV. They proposed that the stretched side of the rod had a positive potential, producing a reverse
bias with the Schottky junction so that no current could flow to screen the polarisation. As the tip left the compressed, negatively polarised side of the rod the junction
was forward biased and therefore current could flow to screen the polarisation.
However, in this work the open-circuit voltage was monitored, and not the current.

This reflects a common confusion in the measurement of this type of device where explanations of current flow are used when measuring open-circuit voltage, when no current can flow in the circuit. The importance of such measurements will be discussed in Sect. 3.3. This explanation did establish the importance of using some kind of potential barrier at at least one contact with the nanorods, which is common with almost all future designs of nanogenerators, and was more fully understood as the significance of screening of the polarisation became apparent, which is discussed in Sect. 3.2.5.

These results were replicated using nanorods grown using solution methods on polyimide (Kapton) substrates [37]. This is the first example of the common zinc nitrate—HMT growth method being used in a piezoelectric application (see Sect. 3.1), which came to dominate the synthesis in later devices. In this study devices were also tested where the nanorods were mechanically stabilised by spin-coating a poly(methyl methacrylate) (PMMA) solution between the rods, a method that also became common in later devices. Although the flexibility of the substrate was not exploited as the measurements were again performed using a conductive AFM tip, this perhaps laid the groundwork for later device designs utilising the flexibility of plastic substrates (Sect. 3.2.3). Apart from a slightly higher voltage output of ~15 mV, this work differed from the original report only in the substrate and growth method used.

As a result of the early experimental measurements into the piezoelectric voltage output of a strained nanorod, a theoretical exploration of this system began to be developed. Gao et al. calculated the potential difference that would develop across a bent nanorod (based on the early AFM tests), which indicated that the main potential difference would develop across the lateral portion of the rod, with an inversion of the polarisation at the constrained base [38]. However, this model neglected the influence of free carriers, which was later included giving a modified potential distribution due to screening by free carriers [39], which is discussed further in Sect. 3.2.5. Importantly, the more realistic inclusion of free carriers gave a more asymmetric potential distribution (see Fig. 3.3), which may account for the ability to measure an external potential difference in an array of ZnO nanorods despite the deformation varying greatly between rods due to differences in alignment.

From the perspective of energy harvesting it is clear that a device that requires the piezoelectric material to be strained by an AFM tip is not practical for widespread implementation. Therefore new ways to replicate this effect were sought, leading to the consideration of methods for straining an entire array of rods simultaneously.

3.2.2   Nanowire Arrays

A key evolution from the early work in measuring voltage output when straining single nanorods was to demonstrate an equivalent effect for entire arrays of nanorods. The first design to achieve this clearly aimed to replicate the structure of an AFM tip, but with an entire array replacing the single tip. This was achieved by creating what

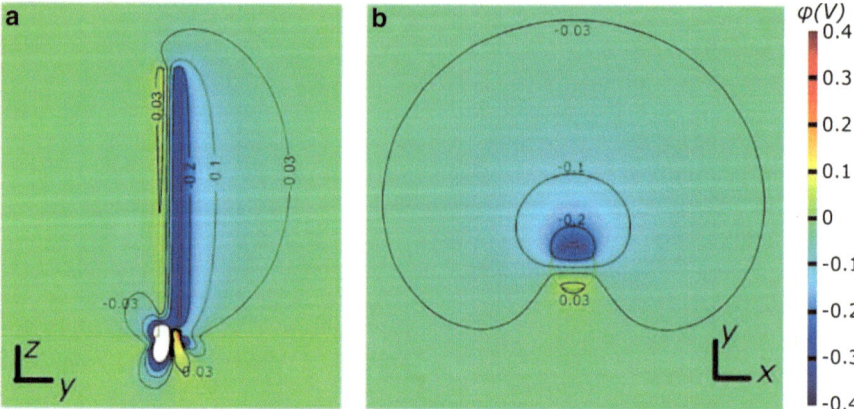

Fig. 3.3 Electric potential (φ) distribution of a bent ZnO nanorod of diameter 25 nm and length 600 nm including the effect of a free electron density of 10^{17} cm^{-3} as modelled by Gao et al. from [39] (rod bending not shown). (**a**) Cross-section of the nanorod length. (**b**) Cross-section across diameter at 400 nm height. The inclusion of free carriers leads to an asymmetrical potential across the diameter, which may account for a measurable potential difference across the length of the rod when bent

Fig. 3.4 SEM micrograph of cross-section of Pt-coated Si 'zig-zag' electrode in contact with an array of ZnO nanorods from [36]

was referred to as a 'zig-zag' electrode, using a patterned silicon surface coated with platinum [36] (see Fig. 3.4). In this case the Pt creates a Schottky barrier with the top of the nanorods, which were grown on either GaN or sapphire substrates. The top electrode was held just in contact with the nanorods by using a polymer spacer around the edge of the device. The entire device was then sealed and placed in an ultrasonic bath so that the top 'zig-zag' electrode oscillated with respect to the nanorods causing them to be strained by the Si 'teeth'. Although this produced a slightly lower voltage output of ~1 mV compared to the AFM tests, attributed to the lower strain induced by the ultrasonic excitation, it did produce a measurable current output up to ~0.5 nA. In these array devices the nanorods are effectively connected in parallel so that the output voltage will be the same as that of a single rod, but the current generated by all the rods is combined and hence will scale with the number of rods in the device, i.e. the device area. Hence it is useful to express the output in units of current per area, which accounts for differently sized devices (see Sect. 3.3). For the given device area in this case, the current density was therefore 25 nA cm^{-2}.

The influence of the early AFM tests persisted for some time with devices still using 'zig-zag' electrodes for a number of years [40], including the use of gold-coated ZnO nanorods as a top electrode [41–44]. These allowed nanowires that were constrained on a rigid substrate to be strained by a top electrode that was often actuated using ultrasonic agitation as in the original report [36]. However, alternative construction methods were also proposed to produce array energy harvesters on rigid substrates. This included the infilling of the nanorods with a polymer material, most commonly PMMA, as first demonstrated by Gao et al. in 2007 (see Sect. 3.2.1) [37]. This was utilised by Xu et al. in 2010, who spin-coated PMMA into an array of chemically grown ZnO nanorods on a gold-coated Si wafer to improve the mechanical stability of the nanorods and to prevent short-circuits to the bottom electrode [45]. In order for the top electrode to contact the rods, the tips were exposed by removing the surface of the PMMA using oxygen plasma etching, before pressing a Pt-coated Si wafer onto the surface. Unlike the device discussed above, these devices were strained by directly pressing with a linear motor. This removed the need for a 'zig-zag' electrode, and also potentially allowed a higher level of strain to be induced in the nanorods, though this was not directly measured. These devices produced a higher open-circuit voltage of 80–100 mV and a short-circuit current density of 4–9 nA cm^{-2}. It was later found that by not removing the surface of the PMMA layer, but instead leaving around 1 µm of PMMA on top of the nanorods before coating the top electrode, a significantly higher voltage could be produced [46]; when Zhu et al. compressed the PMMA-coated ZnO nanorod array grown on Si with a linear motor, they produced an open-circuit voltage up to 37 V after rectification through a bridge rectifier and a short-circuit current density of 12 µA cm^{-2}. They connected a number of devices in parallel and used this to charge a capacitor to store the harvested energy.

Alongside the investigation of nanorod arrays grown on rigid substrates which could harvest the energy from compression, alternative device designs were develop which could be strained through bending as a potential method for extracting a large amount of energy from a range of situations. These devices were facilitated by the use of flexible substrates and are discussed in the next section.

3.2.3   The Introduction of Flexibility

The first example of a nanostructured energy harvester produced on a flexible substrate and tested as a complete array was produced by Choi et al. in 2009 [22]. As with Gao et al. in 2007 (see Sect. 3.2.1) [37], ZnO nanorods were grown at low temperature using a chemical bath of zinc nitrate and HMT (see Sect. 3.1), which was compatible with the flexible conductive indium–tin oxide (ITO)-coated polyethersulfone (PES) substrates. The devices were completed by pressing a top electrode of ITO/PES with or without a Pd–Au coating onto the nanorod array. The authors found that the highest output was achieved with the Pd–Au-coated top electrode, giving 10 µA cm^{-2} when compressed with 0.9 kgf. The improved output with the Pd–Au top electrode was attributed to the presence of a Schottky barrier, as in previous work. Unfortunately the open-circuit voltage was not tested and so full comparison to alternative designs cannot be made. Although these devices were tested as arrays grown on flexible substrates, the impact of flexibility on the output was not evident as simple compression testing was still performed. However, in later devices where alternative top electrodes of either carbon nanotubes [24] or graphene [23] were investigated, the same group performed testing where the entire device was subjected to bending. These produced slightly lower current density outputs of 5 and 2 µA cm^{-2}, respectively which is likely due to the lower conductivity of the carbon-based electrodes introducing series resistance losses.

Following from these early reports, nanostructured energy harvesters produced on flexible substrates became increasingly common. A device very similar to that reported by Gao et al. in 2007 was produced by Lee et al. in 2011 with chemically grown ZnO nanorods on a Kapton substrate [47]. These nanorods were filled with PMMA, exposed using plasma etching and a gold-coated plastic substrate was pressed on top as a counter electrode. A single device produced an open-circuit voltage up to ~350 mV and short-circuit current density up to ~125 nA cm^{-2} when subjected to periodic bending. The authors stacked ten devices on top of each other and connected them together (although the exact details of how this was achieved were not included), a concept previously demonstrated for non-flexible devices [48], increasing the voltage output to 2.1 V. As with the rigid substrates (see Sect. 3.2.2) a similar design where the nanorod surface was left covered by PMMA produced a significantly higher voltage output of 10 V [49]. An additional innovation in this design was that the polyester substrate was coated with ZnO nanostructures on both sides with opposite Au electrodes evaporated on top of each PMMA-coated ZnO layer (see Fig. 3.5). The authors propose that this leads to opposite strains experienced by the nanorod tips at each electrode during bending, increasing the voltage output. The devices produced a short-circuit current density of 0.6 µA cm^{-2}. This demonstrates a common issue with polymer-coated ZnO nanorod devices that generate a high open-circuit voltage: they often produce a low short-circuit current, which will be discussed further in Sect. 3.3.

As demonstrated by the range of device designs discussed above, the majority of ZnO nanorod energy harvesting devices incorporate a Schottky barrier by using a metal top electrode such as Au or Pt on the top of the ZnO nanorods, with or without an intermediate polymer layer such as PMMA. This design follows from the discussion in the original work by Wang and Song (see Sect. 3.2.1) of the importance of a Schottky barrier at the top contact with the ZnO nanorods to generate a power output. The reasons that this barrier may be required are discussed in Sect. 3.2.5, but there have also been alternative designs to the ZnO nanorod array/(polymer)/ Schottky barrier type of device. One such alternative replaced the Schottky barrier with a p–n junction. This was possible because ZnO is most commonly n-type when

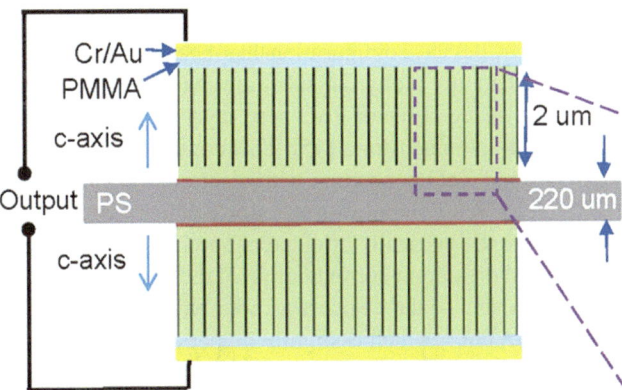

Fig. 3.5 Schematic of double-sided energy harvester with PMMA-coated ZnO nanorod arrays on both sides of a polystyrene (PS) substrate coated with chromium-gold electrodes. From [49]

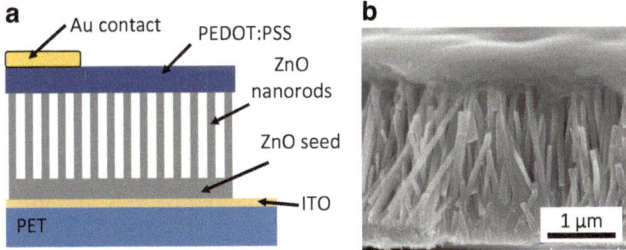

Fig. 3.6 Schematic (**a**) and SEM micrograph (**b**) of cross-section of ZnO nanorod p–n junction energy harvester from [51]

produced due to a range of native defects [50], and therefore a p–n junction was produced by coating the ZnO nanorods with the p-type polymer poly(3,4-ethylene-dioxythiophene) poly(styrenesulfonate) (PEDOT:PSS) [51]. The nanorods were grown on flexible ITO-coated polyethylene terephthalate (PET) substrates, and a gold contact was coated on top of the PEDOT:PSS layer. The PEDOT:PSS layer did not completely fill between the nanorods, unlike previous PMMA-filled nanorods, and instead formed a layer that penetrated just below the ZnO tips, and covered them by ~1 μm. This prevented the gold top contact from short-circuiting to the nanorod tips (see Fig. 3.6). The devices were bent manually giving maximum open-circuit voltage and short-circuit current density outputs of approximately 10 mV and 13 μA cm^{-2}. In a later study of similar devices an open-circuit voltage of 90 mV was produced, and a current density of ~0.5 mA cm^{-2} was measured across an optimum load [52] (see Sect. 3.3.2 for discussion of impedance matching). Although the voltage output was relatively low compared to some PMMA-coated Schottky barrier-type devices, the current density was relatively high, which may be due to the semiconducting nature of the PEDOT:PSS layer compared to the insulating PMMA layer used in many devices.

Although ZnO nanorods generally grow preferentially on many substrate surfaces, an alternative device configuration has been demonstrated where ZnO nanorods are aligned laterally along the substrate. Xu et al. produced this type of device by partially covering strips of ZnO seed layer so that nanorods grew from their sides across the Kapton substrate [45]. This complex lithographic fabrication led to a series of rows of ZnO nanowires with electrodes at each end. By bending this device 1.2 V open-circuit output was produced, which was significantly higher than the ~100 mV from the vertically oriented nanorod device reported in the same paper. The maximum short-circuit current output of the device was given as ~25 nA, and although the exact area of the device was not given, the current density can be estimated around 15 nA cm^{-2}, which is slightly higher than the current density from the vertically aligned nanorod device. The high output voltage can be attributed to the fact that the rows of nanorods were connected in series, and therefore the voltage from each row added to the others. This therefore demonstrates a compact method for adding the voltage from a number of ZnO nanorod arrays in series, though it does require complex lithography. A simpler lateral device was fabricated by Zhu et al. in 2010 by transferring ZnO nanorods from an array to a flexible substrate by horizontal wiping [53]. In a similar way to Xu et al. stripes of Au electrodes were defined lithographically, but the device was simpler as the striped seed layer did not have to be patterned and partially covered. The functioning of the device relies on the fact that the c-axes of the ZnO nanorods were aligned with the same orientation vertically in the original array and are therefore aligned horizontally after transfer. One drawback in this method is that a larger number of rods are not contacted at both ends, as it relies on the Au stripes coinciding with each end of a rod. When bent the device generated an open-circuit voltage up to 2 V and short-circuit current density of 100 nA cm^{-2}.

A further variation for a flexible ZnO nanorod array device is to grow the nanorods onto fibres or cloth or paper substrates. These are extremely inexpensive and very flexible so can produce a large amount of bending with only a small input force. ZnO nanorods grown on fibres for this application were first demonstrated by Qin et al. in 2008 [41]. They grew ZnO on a Kevlar fibre by first sputtering it with a ZnO film to act as a seed layer and then placing it in a solution of zinc nitrate and HMT to grow the nanorods. Tetraethoxysilane was coated between the nanorods in order to improve the mechanical stability and adhesion to the fibre. To strain the nanorods a second fibre covered in gold-coated ZnO nanorods was wrapped around, and the two fibres were stretched across each other (see Fig. 3.7). This produced an open-circuit voltage around 1 mV and a short-circuit current of only 5 pA. The extremely small current in this case may be due to the lack of an electrode below the ZnO nanorod array, as the authors relied on the ZnO seed layer to provide a contact to the base of the rods. This would give the structure an extremely high internal resistance, which would lead to very little current flowing in the external circuit (see Sect. 3.3). This issue remained when work was later published in 2013 with the same ZnO nanorod-coated Kevlar fibres, but overlapping perpendicularly rather than twisted together [54], which still lacked a dedicated conductive electrodes below the nanorods, and only gave a few pA of current. An alternative device using nanorods grown on the fibres of a paper substrate suffered from similar issues, with

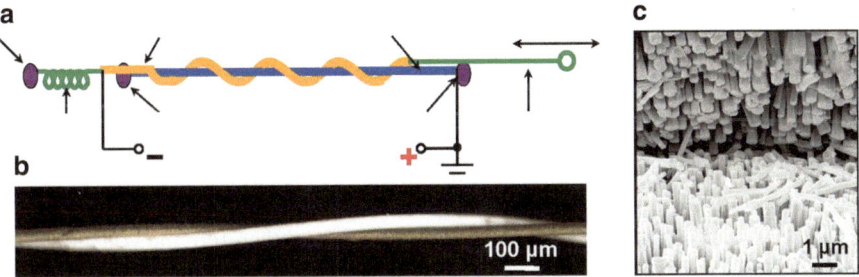

Fig. 3.7 Fibre-based energy harvester using ZnO nanorods coated onto Kevlar fibres. (**a**) Schematic and (**b**) photograph showing interwoven ZnO-coated Kevlar fibres with one ZnO array coated in gold. (**c**) SEM micrograph of facing arrays of ZnO nanorods. From [41]

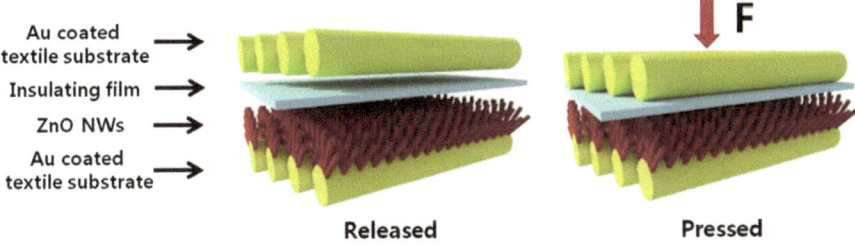

Fig. 3.8 Schematic of ZnO nanorod energy harvester using ZnO nanorods grown on a gold-coated polyester textile substrate with a polyethylene spacer and gold-coated polyester top electrode. From [25]

no conductive electrode below the nanorods, and in this case contacts were only made at each end of the device before the entire structure was bent, generating 15 mV and 10 nA [55]. Khan et al. produced a device on a cloth substrate which did have a full Ag electrode coated on the surface before ZnO nanorods were grown using zinc nitrate and HMT [56]. Strangely they only tested the piezoelectric output by deflecting individual nanorods with an AFM tip and therefore did not assess the performance of a complete device by adding a top electrode and bending it as a whole. In this case they measured 10 mV—similar to the original work by Wang and Song—but did not measure current output. Thus, although the above work has demonstrated that ZnO nanorods can be successfully grown on a range of fibre and cloth-based substrates, they have not integrated this into a successful energy harvesting design that can extract useful power from movement. This, however, has been achieved by Kim et al. who grew ZnO nanorods on Au-coated woven polyester substrates [25]. Not only did this mean that they had a conductive bottom electrodes, but they then completed the device forming a very similar structure to some devices described in Sects. 3.2.2 and 3.2.3 where a gold-coated polyester top electrode was pressed onto the top of the nanorod array, with a 40 µm thick polyethylene (PE) spacer between the nanorods and top electrode (see Fig. 3.8). This generated an open-circuit voltage of 4 V, with a short-circuit current density of 0.15 µA cm^{-2}

when subjected to acoustic vibrations at ~100 dB. The authors also test devices with either only the nanorods, or only the polymer film between the gold-coated fabric and find that 0.5 V and 0.8 V are generated respectively. The authors propose that the voltage generated when only the PE layer is placed between the Au-coated textile originates from an electrostatic generation mechanism because the PE layer has an initial surface charge of -2×10^{-4} cm^{-2}. In this way the device acts similarly to an electret generator, where a material with embedded charge oscillates between two electrodes leading to a variation in the capacitance and therefore the voltage of the system [57]. Therefore the authors propose that the device with both nanorods and a PE layer combine piezoelectric and electrostatic energy harvesting to produce the high output voltage. The ability of this device to generate such a high output from acoustic vibrations is extremely useful as it allows energy to be harvested from noisy environments as well as movement. The authors confirmed that this was enabled by the flexibility of the textile substrate used, as they compared equivalent devices on silicon and polyethersulfone substrates, which generated around one fifth and one half of the output of the textile-based device.

3.2.4 Alternative Materials

Although ZnO nanorods and wires have dominated the field of nanostructured energy harvesters, partly due to the influence of the initial work by Wang and partly due to their simplicity of synthesis, a number of other nanostructured piezoelectric materials have been demonstrated for this application, including materials that have long been used in micro- and macro-scale energy harvesters. In addition, a wider range of nanostructures of ZnO have been investigated with some promising results.

The main example of alternative ZnO nanostructures involves ZnO grown using the same zinc nitrate—HMT chemical method as the nanorods, but which formed nanosheets due to the reaction with the aluminium-coated polyethersulfone (PES) substrate [58] (see Fig. 3.9). By pressing a gold-coated polymer substrate into the nanosheet layer, outputs of up to 0.75 V and 16 µA cm^{-2} were produced in open and short-circuit at a measured force of 4 kgf. The authors highlight the DC (i.e. single polarity) of the output linking it to a 'layered double hydroxide' structure of Zn and Al between the nanosheets and Al electrode. However, such DC-type output is a common feature of nanogenerators using free-standing (or pressed on) top electrodes, and as such appears to be linked to a loss of contact during the release portion of the testing so that the stored charge must dissipate internally rather than through the external circuit—as in the original tests using AFM tips. Another variation of the use of ZnO nanostructures in energy harvesters is to combine ZnO nanoparticles with carbon nanotubes (CNTs). In a limited study these materials were bonded together by cross-linking with benzoquinone to form a ZnO-embedded CNT paper [59]. Electrodes were connected using silver DAG and the paper was bent manually which gave a very small output of ~15 mV. ZnO nanoparticles were

Fig. 3.9 SEM micrograph of
ZnO nanosheets produced
due to the interaction
between the zinc nitrate–
hexamethylenetetramine
synthesis chemicals and the
aluminium substrate. From
[58]

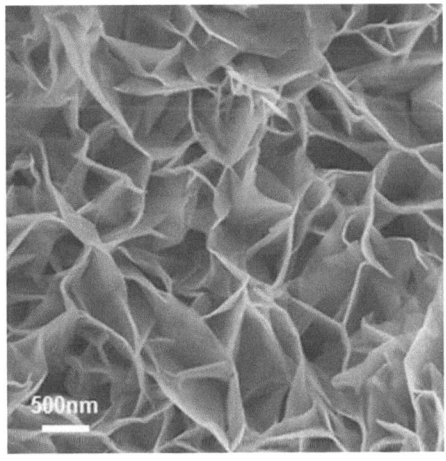

combined more successfully with CNTs by bonding the composite together using
polydimethylsiloxane (PDMS) [60]. Rather than relying on conduction through the
percolated network of CNTs, the authors coated the mixed materials onto a PET/
ITO (indium–tin oxide) electrode substrate and attached a similar substrate on top
of the film. The device could be bent, producing 0.4 V and 50 nA an open-circuit
voltage and short-circuit current, and by manually pressing the film 7.5 V and
2.5 μA were generated. However to demonstrate the durability of the device, the
authors also stamped on it, giving a peak output up to 30 V. This demonstrates how
the output of a piezoelectric energy harvester depends strongly on both the magnitude of force applied and the rate so that almost arbitrarily large outputs can be
produced if high strain rates can be endured [52]. This will be discussed further in
Sect. 3.3. Attempts have also been made to combine ZnO nanorods with the piezoelectric polymer polyvinylidene fluoride (PVDF), which was in-filled between an
array of ZnO nanorods grown on a gold-coated polymer substrate, also coated on
top with gold [61]. When tested during bending the device generated 0.2 V and
10 μA cm^{-2}, which was only slightly higher than the 0.18 V and 5 μA cm^{-2} for a
device without ZnO nanorods, demonstrating that the majority of the power generation originated from the PVDF layer.

Although ZnO-related devices still dominate reports of nanostructured piezoelectric energy harvesters, a small number of studies have been made on alternative
nanostructured piezoelectric materials. 'Nanogenerators' based on barium titanate
(BTO) structures have been reported in 2010 [62] and 2013 [63]. However, the former device did not comprise of nanostructured component, but instead resembled
closely previously studied micro electro mechanical systems (MEMS), where a
BTO thin film with ~50 μm features was deposited on a Si substrate, etched off and
transferred to a flexible Kapton substrate. Although the size of the features and
fabrication procedure of this device make its categorisation as a 'nanogenerator'

Fig. 3.10 Cross-section
SEM micrograph with inset
showing top-down view of
barium titanate nanorods
produced by hydrothermal
conversion of a titanium
dioxide nanorod array. From
[31]

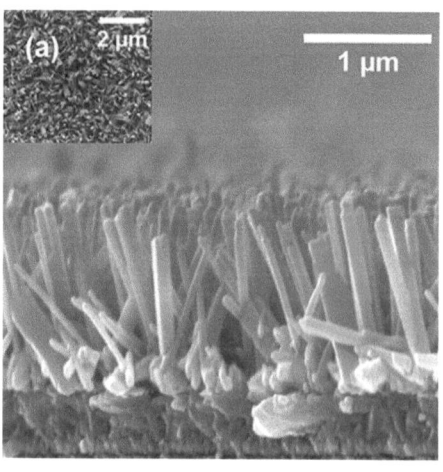

debatable, it does share some features that may enable similar applications such as
the ability to mount the device onto a flexible substrate and the ability of the device
to flex to a large degree without fracture of the piezoelectric film. The latter report
[63] similarly contains a BTO film produced using top-down patterning and etching
techniques, which was not transferred to a flexible substrate in this case. However,
the device was tested more thoroughly than many nanogenerators, where output was
quantified across a range of load resistances (see Sect. 3.3), giving a maximum output of $0.858~\mu W~cm^{-2}$ across a 5 MΩ load. A similar method to the BTO report from
2010 [62] has been used to produce PZT microstructures that were similarly transferred to PET substrates to use as energy harvesters [64]. Again, although the device
was labelled a 'nanogenerator', it comprised solely of microscale features, and the
fabrication resembled more the methods used for MEMS devices. Although the
authors promote the use of top-down methods to produce highly ordered microstructures, it seems likely that such time and energy-intensive methods will remain
more practicable for micro-scale applications, and bottom-up techniques will be
more suitable for covering large areas with nanostructured films suitable for providing higher levels of portable power. Alternatively, a 'nanogenerator' was also
reported using the piezoelectric material $ZnSnO_3$, but used only a single micro-belt
of the material laid onto a plastic substrate and is therefore neither nanostructured,
nor can be envisaged to have significant applications in energy harvesting due to
very low output power [65].

The first energy harvester using BTO that can unequivocally be called a nanogenerator was demonstrated in 2014 by Koka, Zhou and Sodano [31]. The BTO nanorods
were produced by converting hydrothermally grown TiO_2 nanorods grown on conductive fluorine-doped tin oxide (FTO) substrates. The conversion was achieved
using a second hydrothermal process with barium hydroxide, and the nanorods were
poled by applying a high electric field between the device electrodes. Each hydrothermal process was performed at below 250 °C, and both types of nanorods were
1 µm long and 90 nm wide (see Fig. 3.10). The devices were tested much more rigor-
ously than many reports of nanogenerator devices: they were accelerated using a

shaker table at a controlled rate of $1\,g$ (9.8 ms^{-2}) and both the peak-to-peak open-circuit voltage (V_{pp}), short-circuit current (I_{pp}) and maximum power across a matched load were measured (see Sect. 3.3). Although the use of such controlled acceleration meant that fairly moderate 'headline' voltages were generated compared to reports where extremely high impact forces are applied, the report compares equivalent BTO and ZnO directly, allowing the BTO devices to be placed in comparison to the more ubiquitous ZnO. The devices produced outputs of V_{pp}=623 mV, J_{pp} (using quoted device area)=9 nA cm^{-2} and a power density of 6.27 µW cm^{-3} on a 120 MΩ load for BTO and V_{pp}=85 mV, J_{pp}=1.58 nA cm^{-2} and a power density of 0.4 µW cm^{-3} on a 50 MΩ load for ZnO. Hence the BTO devices produced around 16 times more power than the ZnO, which was attributed to the higher electromechanical coupling coefficients of BTO and suggests that BTO-based devices may show promise to produce higher output power than ZnO.

The majority of non-ZnO-based nanogenerators are based on lead zirconate titanate (PZT), which has been widely used in macro- and micro-scale piezoelectric energy harvesters [66, 67]. Early nanogenerators using PZT were very similar to the lateral ZnO energy harvester reported by Zhu et al. [53] described in Sect. 3.2.3: electrospun PZT nanowires approximately 60 nm in diameter and 500 µm long were transferred onto platinum interdigitated electrodes on a silicon substrate and then poled to align their polarisation [27]. This process is necessary due to the ferroelectric property of PZT, unlike ZnO, which is piezoelectric but not ferroelectric and is crystographically aligned when grown directly on a seeded substrate. After encapsulating in PDMS, the structure was compressed with a polytetrafluoroethylene (PTFE) block controlled using a dynamic mechanical analyser. The authors chose to use PZT nanowires as they have previously shown them to have higher piezoelectric coefficients than bulk PZT [68]. The analysis of the output from this device was more thorough than many reports of ZnO nanogenerators, measuring at a range of peak strains and strain frequencies as well as across a range of loads (see Sect. 3.3). The peak output at maximum strain was found at close to 40 Hz excitation frequency, giving 1.42 V peak-to-peak open circuit voltage. The authors calculated the average power across a load for each excitation cycle, obtaining a maximum of 0.03 µW across a 6 MΩ Load. The area of the device was not given, so this could not be scaled per area or volume.

Electrospun PZT nanowires have been used in a number of other nanogenerator devices. Cui et al. expected a higher output than ZnO using electrospun PZT nanowires due to the higher d_{33} coefficient of PZT (500–600 pC/N) compared to ZnO (~12 pC/n) [69]. They confirmed the electrospun fibres were crystalline perovskite PZT using x-ray diffraction. The device consisted of aligned fibres transferred onto a PDMS-coated magnetite (Fe_3O_4) substrate, contacted at each end with silver electrodes, and encapsulated with PDMS. After poling of the PZT the device was bent by moving a magnet below the structure to repel the magnetite substrate generating peak open-circuit voltage values of 3.2 V and open-circuit current of 50 nA. Although strained via a somewhat convoluted non-contact method, this device does demonstrate that nanostructuring can allow materials that are commonly considered brittle to undergo large degrees of strain induced through macroscopic flexing of a substrate, as demonstrated on the microscale in MEMS-based cantilever energy harvesters using ceramic materials [66]. Wu et al. demonstrated that a freestanding, aligned electrospun array of PZT nanowires could be produced by electrospinning onto a series of bridge-like electrodes [28]. This film was bonded to a PET substrate using PDMS and silver electrodes were pasted onto each end. When bent, the device produced up to 6 V in open-circuit and 45 nA of short-circuit current. The device output was measured on a 100 MΩ load, and the average output power was measured to be 0.12 μW, which was calculated to be 200 μW cm^{-3} based on the volume of the PZT portion only (excluding the substrate). It was also demonstrated the PZT array could be lifted off of the substrate to form an extremely flexible (fabric-like) film, which generated lower open-circuit voltage and short-circuit current of 0.24 V and 2.5 nA. Gu et al. reported a method to convert a layered structure of lateral electrospun PZT fibres into a perpendicular array by cutting sections of the film, rotating, and stacking a number of layers, although very sparse details of the process were given [29]. The authors report up to 209 V open-circuit voltage from the device, though they state that this was achieved using a 'larger impact' than was otherwise reported in the paper. As discussed in Sect. 3.3, it is possible to produce an arbitrarily large voltage peak from a piezoelectric energy harvester by applying an extremely rapid and high intensity force, thus this peak value is largely meaningless without quantification of the impact. Using a more controlled impact the voltage does not exceed 10 V.

A hydrothermally grown array of PZT nanorods has also been used in a nanogenerator. As discussed in Sect. 3.1.2, the rods were produced by heating aqueous chemical precursors in a pressure vessel at 230 °C for 12 h [32]. The nanorods grew on epitaxially matched doped strontium titanate substrates, and top electrodes of Ti/Pt on silicon were pressed onto the top surface. By applying a pressure to the top surface voltage peaks up to 0.7 V and current density up to 4 μA cm^{-2} were produced.

PZT has also been used in combination with ZnO nanorods by No et al., where PZT was sputtered onto ZnO nanorods grown using zinc nitrate and HMT on ITO-coated glass substrates and then poled. A Ti/Pt-coated glass substrate was pressed into the surface of either ZnO nanorods only, a PZT thin-film only, or PZT-coated ZnO [17]. These devices gave short-circuit current outputs of 0.75 nA, 7 nA and ~300 nA, respectively. Although the authors suggest the ZnO and PZT outputs are combined in the composite nanogenerator, they do not consider the potential additional screening effect of the PZT on the ZnO, which will be discussed in the following section. Also, without open-circuit voltage measurements it is difficult to compare the devices with other ZnO-only devices in the literature.

3.2.5 The Significance of Screening

As discussed above, all nanogenerator designs to date using ZnO nanorods have incorporated some form of potential barrier at one contact, such as a metal–semiconductor or metal–insulator–semiconductor Schottky barrier or a semiconductor p–n junction.

These potential barriers are normally avoided in energy harvesting devices as a source of potential loss, so to understand why such a barrier is required it is necessary to understand the role of screening in piezoelectric materials.

Screening is well understood and widely studied for ferroelectric materials (see Sect. 2.5), which contain a permanent electrical polarisation controlled by poling; this polarisation originates from the same offset of charged ions in the material as the transient polarisation induced in a piezoelectric by strain. It is the transient nature of the polarisation in a piezoelectric material that is important to consider, especially for energy harvesting applications. The strain-induced polarisation in a piezoelectric produces an internal electric field, which has been described extensively for ferroelectric materials and is known as the depolarisation field [70–72] (see Sect. 2.5). It is this electric field that is measured as a voltage in an external circuit, and which can drive an external current. However, this electric field will also lead to movement of any free carriers within the piezoelectric material, which is related to its finite internal resistance and is known as internal screening [73–76]. Additionally, the polarisation can lead to rearrangement of carriers within the contacts, known as external screening [70–72]. The interactions between this polarisation and metal or semiconductor contacts [70, 71, 77, 78] and the effect of the polarisation on the carriers and band structure of a ferroelectric semiconductor [73–76, 79] have been widely studied for ferroelectric materials.

In a strained piezoelectric, such as in an energy harvester, it is the screening which causes the measured voltage to drop to zero, even if that strain is maintained. As discussed, this screening can originate from internal carriers or the external contacts. However, it has been calculated that due to the small size of the nanostructures involved and the drift velocity of the carriers, the internal screening occurs on timescales faster than any measured voltage output (hundreds of nanoseconds) [51]. Therefore any voltage output from a nanogenerator represents the net field from the polarisation already screened by internal carriers. It has been demonstrated that this is feasible as it has been calculated that at carrier densities below 10^{18} cm^{-3} the polarisation cannot be completely screened [80], and carrier density of chemically synthesised ZnO has been measured as 1.3×10^{17} cm^{-3} [81]. The effect of varying the carrier density of ZnO dynamically has been demonstrated by illuminating with UV light photoexcitation, showing a clear drop in voltage output when illuminated [40, 51, 82]. Therefore, to increase the voltage output from the devices the carrier density of the ZnO should be reduced, which is discussed below. However, it also indicates that the time-dependent reduction in voltage output—clearly visible when measurements are taken with high time resolution [52]—is a result of the external screening. It is here that the key lies to the need for non-Ohmic contacts to nanogenerator devices: in Ohmic metal contact the density and mobility of screening charges is extremely high [71], therefore the polarisation will be screened so rapidly that no external voltage will be measured [51]. However, if the contact incorporates a Schottky or p–n junction a depletion region will exist, which lowers both the density and mobility of the carriers due to the built-in field at the junction [51]. It is for this reason that the nanogenerators only operate successfully with a Schottky contact or p–n junction rather than Ohmic contacts [45]. Thinking of the device in terms of an equivalent circuit model, the screening charges can be thought of as charging the internal capacitance of the energy harvester, which opposes the voltage generated in the piezoelectric material [83]. Thus, even in open circuit when a net current cannot flow in the circuit, the redistribution of charges can cause the externally measured voltage to drop to zero. In this case the addition of the junction to the structure can be thought of as altering the time constant of the system, which is discussed further in Sect. 3.3.3.

As the use of non-Ohmic (generally Schottky) contacts has been well-established since the inception of piezoelectric nanogenerators, the focus for the improvement of voltage output though the reduction in screening has been on reducing the internal free carriers from the levels found in as-synthesised ZnO. These fall largely into two categories: 'bulk' effects, whereby the defect-induced carrier density is reduced by treating the entire of the ZnO material, and surface effects, where surface treatments are used to reduce the free carriers induced by surface states such as adsorbed molecules and dangling bonds.

One method to reduce the carrier density of chemically synthesised ZnO nanostructures is through heat treatment. This can reduce the density of intrinsic defects resulting from the low-temperature synthesis method and potentially remove any adsorbed molecules remaining from the chemical precursor solution. This method was used by Hu et al. to increase the output of a PMMA-filled ZnO nanorod array on a polymer substrate from around 2.5 V and 150 nA cm^{-2} before treatment to around 8 V and 900 nA cm^{-2} after thermal treatment at 350 °C for 30 min in air. Treatment at such high temperature was enabled by the use of Kapton substrates. The authors also used oxygen plasma treatment of the nanorods to reduce surface defects, giving an output of 5 V and 300 nA cm^{-2} which dropped over time due to exposure to air despite the PMMA encapsulation. The most effective treatment in this report was to coat the nanorod surface with a layer-by-layer polyelectrolyte coating of poly (diallyldimethylammonium chloride) (PDADMAC) and poly (sodium 4-styrenesulfonate) (PSS), leading to an output of 20 V and 6 μA cm^{-2}. Although the carrier density of the ZnO nanorods was not measured before or after each treatment, this suggests that the largest source of free carriers in the ZnO, and therefore greatest potential area for improvement, is the surface states present after synthesis. This can be attributed to the large surface-to-volume ratio of the nanorods. Another study using heat treatment of ZnO to increase the device output from ~50 mV to 300 mV also annealed the nanorods in air at 350 °C for 30 min [82]. Significantly the authors also found that after annealing the device output dropped very little under UV illumination (to 250 mV) indicating excitation or trapping in defect states with illumination contributes significantly to the polarisation screening. Such stability to external stimulus is important as energy harvesters need to generate a reliable output in a range of environments, although response to stimuli such as UV illumination could have applications in self-powered sensors, as discussed in Sect. 3.2.6.

Another route to reduce the intrinsic carrier density of as-grown ZnO nanostructures is to introduce dopants that compensate for the intrinsic defects. As the majority of native defects in ZnO are donor-type, these compensating defects should be acceptors.

Such acceptor doping can be achieve by partial substitution of Zn by elements with one less valence electron, which are found in group 1 and group 11 of the periodic table. Both of these groups have been investigated by researchers at Samsung and Haonyang University in Korea, using either lithium [84] or silver [85] doping of solution-grown ZnO nanorods. In the case of Li-doping the ZnO nanorods were grown on silicon substrates and the lithium was incorporated by adding lithium nitrate to the growth solution. In addition to the Li-doping, the surface of the nanorods was also passivated using oleic acid, a long-chain organic acid which can bond to the ZnO surface via the carboxyl group. As the nanorods were grown on rigid substrates they were tested using acoustic vibrations and gave over 20 times more voltage with Li-doping, and 30 times more voltage with additional surface passivation. The maximum peak-to-peak voltage generated by the devices was 2.9 V, which is very large considering that they were excited with only acoustic vibrations. Devices produced using silver-doped ZnO were grown on Au-coated polyester fibres, similarly to the work by Kim et al. discussed above [25]. The Ag-doped ZnO nanorods produced a peak-to-peak voltage of 4 V and current of 1.0 µA when excited with sound at 80 dB, which were around three times greater than the devices using undoped rods.

Finally, the use of p-type layers on the ZnO nanorods could lead to some screening of the surface states, though this is dependent on the degree of coating achieved. Coating with a p-type layer was attempted early in the development of ZnO nanogenerators, where the early work of Wang and Song was replicated using nanorods coated with the p-type oligomer (2,5-bis(octanoxy)-1,4-bis(4-formyl phenylene vinylene) benzene) (OPV2) [86]. However, the authors focussed on the effect of the p–n junction as the significance of screening had not been considered thoroughly at this time. In the case of the ZnO/PEDOT:PSS devices discussed in Sect. 3.2.3, there is little penetration of the p-type layer between the nanorods, so the majority of the rods remain uncoated. Hence in this case the advantage of this p-type material largely results from the reduction of external screening from the contact. As discussed in the previous section a PZT-coated ZnO nanorod device was shown to have increased output compared to an uncoated nanorod device [17]. As well as a potential additional piezoelectric output from the PZT layer it is likely that it will have passivated the ZnO surface. Such combinations of conformal p-type layers with ZnO nanorods therefore potentially have the advantage of reducing both internal and external screening of the polarisation in ZnO, with even the potential to add additional piezoelectric materials such as PZT.

3.2.6   Applications

In the discussion above, the output from the various designs of nanogenerator have been compared mostly based on peak values of open-circuit voltage or short-circuit current. These outputs are generally made up of randomly or regularly oscillating peaks with either a single or dual polarity. Though some single polarity outputs have been described as 'DC' they do not comprise of a continuous, regular output and

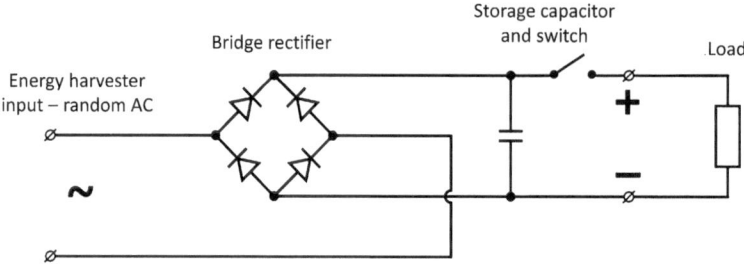

Fig. 3.11 Bridge rectifier and storage circuit for rectifying the random alternating current output of a piezoelectric energy harvester in order to charge a capacitor, which can then be switched in order to power a load

therefore are not directly useable to power a device. The majority of portable devices that may be powered from an energy harvester require a DC supply at a fixed voltage, which would otherwise be supplied from a battery. In fact, if longer term storage of the harvested energy is required, the output from an energy harvester may be required to charge a battery for which a regulated voltage would be necessary.

Some proposed applications of these energy harvesters are as 'self-powered' sensors. In this case variations in the output peaks caused by external stimuli are proposed to be used as measures of those stimuli [45] and therefore charge regulation and storage is not required. For example, as discussed above, the output can vary due to UV illumination; therefore the proposal is to measure the illumination from the variation in voltage output. However, a significant flaw in this proposed application is that as well as varying due to the stimuli to be measured, the output of the device will also vary due to the speed and magnitude of the input force. Therefore, unless the input force is mechanically regulated (and therefore requiring external power) the only application of these devices as self-powered sensors is in measuring only the input force. In all other cases it is necessary to rectify and store the output ready for use in external sensors and transmitters. Such a self-powered force sensor was demonstrated by coating a steel spring with ZnO nanorods and using the variation in bending of the spring to act as a self-powered balance [87].

Many examples of such regulation and storage can be found in reports of nano-generator devices. In the majority of cases a bridge rectifier is used to convert voltage outputs of both polarities to a single polarity. This is generally then used to charge a capacitor to a required voltage, after which a switch is closed connecting the capacitor to a device such as an LED, LCD display or sensor (see Fig. 3.11) to provide a short pulse of power. The use of a switch in the circuit is not commonly highlighted, but it must be noted that this requires the operator to monitor the voltage developed on the capacitor and only close the switch when this is sufficient to power the device. This hands-on operation must of course be removed for truly autonomous operation, such as in a sensor node, but this could be achieved relatively simply using a small amount of logic circuitry. Another factor that should be considered is the time scale of the charge storage. In principle it is possible for any energy harvester that generates a voltage above the turn-on voltage of a device such as an LED to provide it with power given an arbitrary length of time to charge a capacitor. However, for practical applications delivery a few milliseconds of power after hours of charging may not be sufficient, although wireless sensor nodes can potentially operate in this fashion if infrequent measurements are required.

In early examples where nanogenerators were used to power external devices the capability was demonstrated simply by causing an LED to flash after sufficient charge has been stored [32, 53], which has since become a common demonstrator of a nanogenerator's ability to power a device [25, 29, 44, 47]. Beyond these demonstrations other electronic devices have been powered such as LCD displays [25, 28, 85], in some cases powering a display for several seconds after a few minutes of charging [25]. Also, an electrophoretic [84] display was attached to a nanogenerator through a rectifying and storage circuit, which only requires power to switch the display and is therefore a good choice of application for such a short pulsed output.

The capability of nanogenerators for powering a sensor node was demonstrated effectively by Hu et al. in 2011 [49]. They produced a double-sided ZnO nanogenerator on a flexible polyester substrate (discussed in Sect. 3.2.3) with nanorods coated in PMMA before adding gold electrodes on both sides. Assisted by the 10 V peak output of the device, they repeatedly flexed it to charge a capacitor through a bridge rectifier. They then connected a photodetector and radio transmitter to the circuit, which could measure an optical signal and transmit it using the radio transmitter (see Fig. 3.12). Although the low power consumption of the transmitter meant that it could be powered without the photodetector after only three bending cycles of the device, the photodetector had considerably higher power requirements and could only operate after 1,000 bending cycles. This demonstrates the need for the power outputs from nanogenerators to be increased significantly to be able to power useful devices. An alternative to a wireless transmitter to output the signal from a sensor is to use an LED as an indicator. This was demonstrated for a mercury sensor, which operated based on the variation in resistance of a field-effect transistor (FET) when Hg was present in an aqueous solution [47]. This system was powered using a ZnO nanogenerator, with the charge also rectified with a bridge rectifier and stored in a capacitor. When sufficient charge was stored and Hg was present on the sensor, the current flowed from the capacitor and lit the LED, indicating the presence of mercury.

Rather than charging a capacitor, it has been demonstrated that a nanogenerator can be used to charge a battery. To enable this, a circuit using a commercial energy harvesting chip (LTC3588-1, Linear Technology) was connected to the nanogenerator and used to charge a watch battery [88]. After straining the device for 20 min the watch was powered for 1 min. Again, if the power (i.e. charge) output from a nanogenerator can be maximised then applications such as charging a battery from movement or vibrations will become more feasible. This energy harvesting circuit has been compared with a standard bridge rectifier and a synchronous switched harvesting on inductor (SSHI) system for use with a nanogenerator [89]. Since many nanogenerators do not produce enough voltage and/or current to operate an external device such investigation of how they will be used in actual applications are rare.

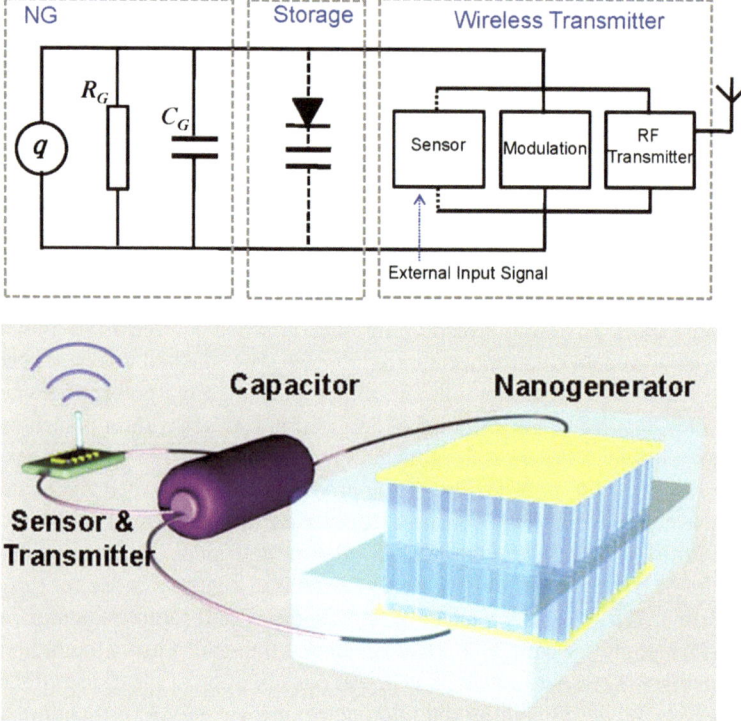

Fig. 3.12 Circuit diagram and schematic of nanogenerator used to power a sensor and wireless transmitter, therefore acting as a wireless sensor node. From [49]

However, as the output of nanogenerators increase to useful levels it will become increasingly important to consider their suitability for energy harvesting circuits optimised for MEMS-based and macroscopic energy harvesters. This study found that the SSHI system required too much current to transfer any charge to a storage capacitor, even though it should have solved the issues with the bridge rectifier of lost power during the bending cycle due to reverse biasing of the diodes from the partially charged capacitor. Thus it was found that the LTC3588 circuit produced the most efficient method to rectify and store the output from a nanogenerator, supplying around twice the current of the bridge rectifier.

3.3   Testing and Performance

In the previous section the output produced by a range of nanogenerators was reported, and most commonly open-circuit voltage and/or short-circuit current or current density values were given. However, great care should be taken when comparing these values, as there are a number of factors that should be taken into account about the testing and reporting of energy harvesters and specifically nanogenerators. These relate to both the way that the output was generated—i.e. the method used for straining the nanogenerator—how it was measured, and what value is chosen to be reported. Although these aspects are commonly taken into account for micro- and macro-scale energy harvesters, they are not routinely considered in the characterisation of nanogenerators. The significance of these aspects of testing and reporting of performance, and the methods commonly used for testing nano-generators, are considered in this section.

3.3.1 Mechanical Input: Methods and Characterisation of Device Straining

The early measurements of voltage output from strained ZnO nanorods were achieved by deflecting single rods using an AFM tip. The tip was coated in platinum so that it was conductive, and the tip and conductive substrate on which the nanorods were grown were connected to the measurement system. This conductive AFM setup allows electrical measurements to be made through the nanorods or in the case of piezoelectric transduction for any electrical output from the nanorods to be detected. The AFM tip thus serves the dual role of inducing the bending of the nanorods and measuring the output. However, potential issues have been identified with this methodology surrounding the fact that contact is made and then lost simul-taneously with the measurement of the output [34]. This will lead to an abrupt change in resistance in the measurement circuit as the tip contacts the relatively low resistance ZnO nanorods and then releases them returning to open-circuit condi-tions. Alexe et al. showed that output voltage spikes can be measured when scanning either ZnO or (non-piezoelectric) Si nanorods, suggesting that care must be taken when measuring piezoelectric output signals. The voltage signals detected in the early reports by Wang et al. were also unipolar, which led to claims of DC output. As discussed previously, such reports of unipolar output have been repeated in some later devices, but these devices commonly use separate conductive top contacts, which, like the AFM tip in the original report, are able to lose contact with the ZnO surface after it has been strained. It therefore seems likely that the unipolar output in these cases is a result of a loss of contact with the piezoelectric material after the stress has been released.

When testing of the piezoelectric output from nanostructured material expanded from single-rod testing to testing of nanorod arrays, a much wider range of tech-niques for straining the piezoelectric material became available. Testing of arrays on rigid substrates was largely confined to either acoustic excitation, largely using ultrasonic vibrations, or by compression of the nanostructured array by pressing on the top layer. Where a device is designed to respond to acoustic vibrations, a mea-sure of the intensity of sound such as a dB level is invaluable. The latter method could be achieved either manually, e.g. by pressing with a finger, or mechanically. Pressing manually clearly requires no additional equipment, but is very difficult to control with the rate and extent of strain varying for each press, and no way of measuring how much stress is being applied. Although straining the device using a mechanical actuator such as a linear motor is much more controllable and measurable, the actual rate of strain, acceleration or pressure applied is rarely reported for nanogenerator devices. In order to make reasonable comparison between different devices, at least some of these parameters are essential. This is because the measured voltage and current from a strained piezoelectric depends directly on the magnitude and rate of stress or strain. The dependence on the magnitude can be seen by considering the relationships between electric displacement (D) and stress (T) set out in Sect. 2.2 and the corresponding equation for strain (S) (omitting the tensor notation):

$$D = dT + \varepsilon^T E; \tag{3.1}$$

$$D = eS + \varepsilon^S E. \tag{3.2}$$

Thus, for constant piezoelectric coefficients d or e, a higher magnitude of stress or strain results in a higher electric displacement. Furthermore, current is given by the time derivative of the displacement field. Thus at constant electric field:

$$I = \frac{dD}{dt} = d\frac{dT}{dt}; \tag{3.3}$$

$$I = \frac{dD}{dt} = d\frac{dS}{dt}. \tag{3.4}$$

So that the short-circuit current (or the open-circuit voltage) depends directly on the rate of application of stress and resulting stain rate of the piezoelectric material. Therefore by applying stress more rapidly a higher output peak can be generated. Since nanogenerators are commonly tested with impulse-type excitations, and peak values of current or voltage are reported, it is essential that some measure of the magnitude and rate of stress or strain are given. Although in a complex composite of multiple nanostructured piezoelectric crystals with substrate and contact layers, and commonly in-fill materials such as PMMA, it is almost impossible to quantify the actual degree or rate of strain of the individual piezoelectric elements, full details of the application of stress to the entire system is still valuable for assessing reported output. Even if a well-controlled mechanical system such as a linear motor is not used, or where the movement of the device is not directly proportional to the movement of the actuator, it is still possible to measure the actual motion using a measurement system such a laser triangulation sensor or a Doppler system [52].

The measurement and quantification of the straining of a nanogenerator device becomes more complex as the testing of flexible devices is considered. These are commonly tested by bending the substrates, which is seen as an advantage of such flexible devices as it allows the harvesting of energy from high displacement but low frequency movements, such as human motion, rather than the high frequency, low displacement vibrations commonly targeted by MEMS-based devices [67]. A common method used to bend such devices is to compress them along their longest axis so that they are forced to bend perpendicular to the direction of compression. The exact nature of the strain induced by the deformation is not known, but it is likely that an array of nanostructured piezoelectric material grown on the substrate will experience some bending and some compression as the entire substrate is bent. As discussed above, any peak voltage or current outputs from such method can be increased by increasing the rate of bending, therefore the speed and displacement of the linear motor should be given. In some cases the degree of curvature of the substrate is given, which is also useful. However, in some cases a calculation is presented of the actual strain induced in the piezoelectric elements. These can only be calculated based on a significant number of assumptions and therefore are unlikely to be accurate. Other methods for bending flexible substrates are also possible such as direct deflection of one end using a linear motor while the other is held rigid, or deflection of the end by another means such as a mechanical cam. In all cases it is most useful if an actual displacement profile of the motion is recorded [52], as this allows the maximum acceleration of the sample to be calculated.

3.3.2   Measurement of Nanogenerator Output

As discussed, the most common output parameters reported for nanogenerators are open-circuit voltage (V_{oc}) and/or short-circuit current (I_{sc}). Although these can be a useful metric for comparison between devices if attention is given to the straining methods described in the previous section, there are a number of issues that should be considered with their use. The most common and most significant issue is to report the power output of the device by merely taking the product of these values, i.e. $V_{oc} \times I_{sc}$. These two values are measured in completely different scenarios— either open-circuit or short-circuit conditions, and their product is meaningless as that power value could never be delivered to a real load, such as a sensor, transmitter or storage capacitor or battery. The optimum transfer of power using complex energy harvesting circuits has been considered in detail for other energy harvesting technologies [90], but as mentioned in the previous section these circuits have rarely been tested with nanogenerators [89]. Instead, the most basic test to determine the maximum power output of an energy harvester is to measure the voltage across (or current through) a range of resistive loads connected to the device (see Fig. 3.13). This can be performed simply by connecting the device to a range of single resistors and measuring the voltage or current, or by using a variable resistor, perhaps also controlled using an automated test system. The actual power (P) delivered to the resistor can then be calculated by:

$$P = \frac{V^2}{R},$$
(3.5)

or

$$P = I^2 R,$$
(3.6)

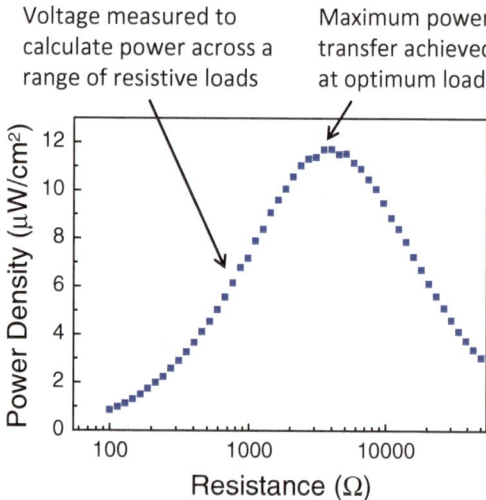

Fig. 3.13 Example load curve showing power calculated from Eq. (3.5) by measuring the voltage generated by an energy harvester across a range of resistive loads. This device produces a maximum power transfer at a load of 3–4 kΩ

using either the voltage, V, or current, I, measured across or through the resistor, R. Such measurements have rarely been reported for nanogenerators [27, 31, 52], but it has been shown that it leads to power values around 3–4 times lower than simply calculating $V_{oc} \times I_{sc}$ [52]. It should also be noted that measuring across a resistive load will not give the maximum power delivered to a reactive load, such as a capacitor, and as such a range of reactive loads could be tested, though this is complicated by the time-dependence of the voltage when testing reactive elements, considering the short duration of output voltage from a nanogenerator.

In addition to using only V_{oc} and I_{sc} values to calculate power, the values reported are nearly always the peak output. This is a consequence of the tendency to test using impulse types of excitation (i.e. single bending cycles) rather than, for example, sinusoidal vibrations. However, consideration should be given to the potential average power generated by the nanogenerator, as this is what can be delivered to an external device. The average power for a regularly oscillating AC source with a sinusoidal output can be calculated relatively simply from the root-mean-square (RMS) voltage and/or current (measured across a load):

$$P_{ave} = I_{RMS} \times V_{RMS} = \frac{V_{peak}}{\sqrt{2}} \times \frac{I_{peak}}{\sqrt{2}} = \frac{V_{peak} \times I_{peak}}{2}. \tag{3.7}$$

Hence the RMS power is one half the power measured from the peak of the output. This represents the maximum proportion of the average power to the peak power, as it originates from a regularly oscillating source (and therefore could potentially be achieved by a device harvesting acoustic vibrations). However, as mentioned, most nanogenerators are strained using impulse-type excitations. Therefore the voltage or current output comprises a sharp spike rather than a sinusoidal oscillation. Thus the average power must be calculated more generally averaging the integrated value over the duration of a pulse. For the case of a voltage measured across a resistor:

$$P_{ave} = \frac{1}{T} \int_{t_1}^{t_2} \frac{V(t)^2}{R} \, dt, \tag{3.8}$$

where t_1 and t_2 are the start and end of the pulse, and T is the duration $(t_2 - t_1)$. In order to calculate this for an arbitrary shape of pulse high time resolution measurement is required, for example using an oscilloscope. It should be noted that if this calculation is performed across a single cycle, each impulse (e.g. bending cycle) must follow immediately from the previous for this average power to be delivered continuously. When the output may only last for milliseconds this implies a deformation frequency of tens to hundreds of Hz, which may not be physically possible, or may lead to mechanical fatigue of the device. Even considering just one cycle, it has been shown that the average power is many orders of magnitude lower than the peak power, which is very important when considering the possible devices that can be powered using these energy harvesters. In addition to the average power, a number of parameters can be calculated for a single cycle, such as the total electrical energy transferred:

$$E = \int_{t_1}^{t_2} \frac{V(t)^2}{R} \, dt, \tag{3.9}$$

and the total charge transferred per cycle:

$$Q = \int_{t_1}^{t_2} I(t) \, dt. \tag{3.10}$$

These can be useful to compare the energy output to energy input—i.e. energy conversion efficiency (though quantification of the input energy must be performed carefully)—or to calculate the maximum charge that could be transferred to a capacitor or battery.

Finally, a short note on the use of current or current density is useful. As most nanogenerators are based on a nanostructured material deposited or grown on a substrate, the current output generally scales with area. Thus the current, and therefore power, can be increased by increasing the size of the device. Hence, as with photovoltaic devices, it is useful to use the current density ($J_{sc} = I_{sc}/area$) when reporting output from devices. Then the power per area can be calculated, again providing a more useful comparison between devices. In addition, many bulk and MEMS-based piezoelectric energy harvesting devices report volume power density (power divided by the volume of the device) [57, 67, 91]. However, there may be different methods to calculate the volume of the device, especially considering

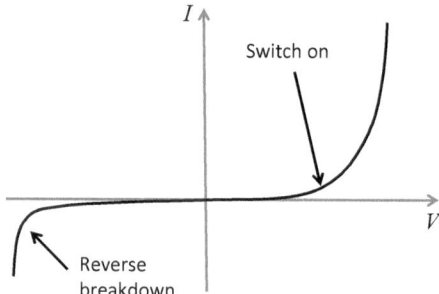

Fig. 3.14 Schematic of typical non-ideal diode current (I)–voltage (V) behaviour for a Schottky or p–n junction showing gradual switch on in forward bias and breakdown at large reverse bias. Normal testing regimes should not take a device to reverse breakdown as it may be irrevocably damaged

whether the substrate and supporting material is included. For nanogenerators it seems common for only the thickness of the *active* area to be used. Although this may be reasonable as the substrate thickness is somewhat arbitrary, some substrate is always required, and neglecting its thickness can lead to artificially high volume power density values compared to other technologies.

3.3.3 Electrical Characterisation

The potential issues suggested for the initial measurements of voltage outputs from ZnO nanorods strained using an AFM tip highlight the need to consider both the methods used to measure the output from the devices and the internal electrical properties of the devices. A range of electrical properties can be measured for nano-generators including current–voltage (I–V) characteristics and impedance measurements. Considering the aforementioned requirement for a barrier such as Schottky or p–n junction, a basic I–V sweep obtained by applying a range of voltages and measuring the current through the device can be useful to determine whether such a junction is present. If it is, some form of rectified (i.e. asymmetrical) I–V characteristics such as those seen in Fig. 3.14 should be obtained.

Impedance analysis is an extremely valuable technique to understand the electrical properties of an energy harvester in more detail. By measuring the variation in impedance of the device when an AC voltage is applied at a range of frequencies, both the resistive impedance (frequency-independent component, Z_{Re}) and reactive impedance (frequency-dependent component, Z_{Im}) can be measured. These can be plotted as the x- and y-axes of a graph to form a Nyquist plot, which is a useful method to visualise the impedance properties of the device. A piezoelectric energy harvester is commonly modelled as an RC circuit, which gives a semi-circular relationship between Z_{Re} and Z_{Im} (see Fig. 3.15). Such measurements are not commonly

Fig. 3.15 Graph showing Nyquist plot of the ideal semi-circular response from a simple *RC* circuit. The value of real (resistive) impedance is equal to the diameter of the curve on the real axis (Z_{Re}), and the value of capacitance can be found from the frequency f_c, where $-Z_{Im}$ is at a maximum. *Inset* shows circuit diagram of the *RC* circuit

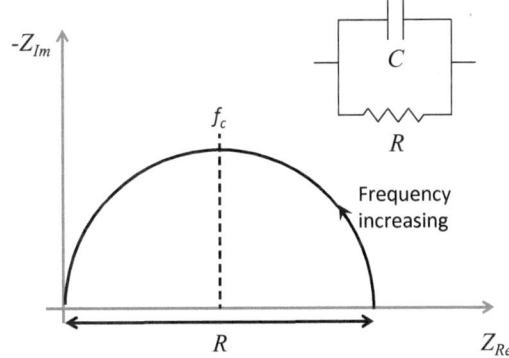

performed for nanogenerator devices, but has been shown to produce such an RC-type response [52]. An excellent example of thorough impedance analysis of a nanogenerator is the BaTiO$_3$ nanorod device discussed in Sect. 3.2.4, which includes resistive and reactive impedance measurements and comparison to an equivalent circuit [31]. As discussed above, both the internal resistive and reactive impedance of the devices should be taken into account when considering output measured using a particular measurement system. If the impedance of the device and that of the measurement system are not dissimilar by orders of magnitude then an accurate measure of the output values will not be obtained. An understanding of the impedance properties of the device will also help to identify the optimum load impedance for maximum power transfer, as this will be close to the internal impedance of the device. By modelling the device as an RC circuit the resistance and capacitance can be obtained from the Nyquist plot using the relationship:

$$RC = \frac{1}{2\pi f_c}. \tag{3.11}$$

Here the resistance, R, is simply the diameter of the semicircle along the real axis of the Nyquist plot, as this gives the frequency-independent component of the impedance. f_c is the frequency at which $-Z_{Im}$ is a maximum, from which the capacitance, C, can also be calculated [92]. This calculation will hold more accurately the more closely the device resembles a simple RC circuit, where the maximum of $-Z_{Im}$ should be half the value of R (i.e. it should be a perfect semicircle).

More detailed consideration of factors that can influence measured output of an energy harvester has been described in depth for non-nanostructured devices, but applies equally to nanogenerators. In addition, significantly more detail and complexity of analysis of impedance spectra is possible. Therefore for further reading on the significance of the measurement system, device properties and impedance analysis readers should look elsewhere [57, 66, 67, 92–94].

3.4 Future Prospects

To date there have been over 100 reports of 'nanogenerator'-type devices, where nanostructured piezoelectric materials have been used to convert mechanical to electrical energy. Some of these have been summarised in this chapter, including a wide range of ZnO nanorod-based devices with peak outputs regularly reaching a number of volts, as well as devices based on nanostructures of other piezoelectric materials such as lead zirconate titanate and barium titanate. These have been produced on both rigid substrates which can respond to pressure and acoustic vibrations and flexible substrates which can transfer large magnitude movements into strain in the piezoelectric materials. With the high output voltages potential applications have been demonstrated, as summarised in Sect. 3.2.6, by rectifying the output from the devices and storing in a capacitor or battery. It is such applications that are the ultimate goal of energy harvesters, and as such it seems likely that the focus of research should lie in this area. In order to achieve this there is a need both to maximise the amount of power generated by the device and the amount of this power that can be usefully transferred to an external device. The maximisation of the generated power can be split into two areas. The first is the optimisation of the conversion of the strain experienced by the piezoelectric material to electrical energy. This essentially equates to the internal efficiency of the material, and will be achieved by optimising the material parameters through synthesis, post-treatment and device construction (Sect. 3.2.5) as well as by using piezoelectric materials with the highest possible electromechanical coupling coefficients and low internal losses (Sect. 3.2.4). The second way that power conversion can be maximised is to design a device to transfer the maximum amount of mechanical energy from the environment to the piezoelectric material. This is dependent on the application, but as discussed previously (Sect. 3.2.3) devices based on flexible substrates have the potential to convert large movements such as body motion into electrical energy. Fabric-based devices are also promising for this application, as well as having been shown to effectively convert acoustic vibrations to electrical energy. The conversion of sonic and ultrasonic vibrations will depend largely on the coupling efficiency of the substrate on which the piezoelectric is grown or deposited, and there is a large potential for maximising the power transfer in this case. A large amount of application-based optimisation has been performed for MEMS-based devices [57, 66, 67], and although some simple real-world type tests have been performed using nanogenerators, a concerted effort to optimise their structural design is still required. In order to quantify any gains made through these areas of development, there is still a need to realistically characterise both the internal electrical properties of the devices, the method used and magnitude of strain applied to the device, and the size of the output, as discussed in Sect. 3.3. Although a minority of nanogenerator reports include thorough characterisation, there is still a tendency to overlook such precise testing and reporting, which must be addressed if devices are to be realistically compared and progress is to be made. Finally, the transfer of power from the nanogenerators to storage media and external devices should be considered. As discussed, there has only been one report to date comparing energy harvesting circuits for use with nanogenerators [89], compared with a large body of work for macroscopic and MEMS-based devices [90]. It is likely that many of these developments will be transferrable to nanogenerator applications, but there is a need to consider the optimum power transfer technology if nanogenerators are to make the transition from the research laboratory to real-world implementation.

References

1. Law M, Goldberger J, Yang P (2004) Semiconductor nanowires and nanotubes. Annu Rev Mater Res 34:83–122
2. Schmidt-Mende L, MacManus-Driscoll JL (2007) ZnO - nanostructures, defects, and devices. Mater Today 10:40–48
3. Yi G-C, Wang C, Park WI (2005) ZnO nanorods: synthesis and characterization and applications. Semiconductor Sci Technol 20:S22–S34
4. Jie J, Wang G, Chen Y, Han X, Wang Q, Xu B, Hou JG (2005) Synthesis and optical properties of well-aligned ZnO nanorod array on an undoped ZnO film. Appl Phys Lett 86:1–3
5. Conley JF Jr, Stecker L, Ono Y (2005) Directed assembly of ZnO nanowires on a Si substrate without a metal catalyst using a patterned ZnO seed layer. Nanotechnology 16:292–296
6. Li C, Fang G, Su F, Li G, Wu X, Zhao X (2006) Synthesis and photoluminescence properties of vertically aligned ZnO nanorod-nanowall junction arrays on a ZnO-coated silicon substrate. Nanotechnology 17:3740–3744
7. Wang L, Zhang X, Zhao S, Zhou G, Zhou Y, Qi J (2005) Synthesis of well-aligned ZnO nanowires by simple physical vapor deposition on c -oriented ZnO thin films without catalysts or additives. Appl Phys Lett 86:24108
8. Lévy-Clément C, Tena-Zaera R, Ryan MA, Katty A, Hodes G (2005) CdSe-sensitized p-CuSCN/nanowire n-ZnO heterojunctions. Adv Mater 17:1512–1515
9. Tena-Zaera R, Katty A, Bastide S, Lévy-Clément C, O'Regan B, Muñoz-Sanjosé V (2005) ZnO/CdTe/CuSCN, a promising heterostructure to act as inorganic eta-solar cell. Thin Solid Films 483:372–377
10. Tena-Zaera R, Ryan MA, Katty A, Hodes G, Bastide S, Levy-Clement C (2006) Fabrication and characterization of ZnO nanowires/CdSe/CuSCN eta-solar cell. Comptes Rendus Chimie 9:717–729. doi:10.1016/j.crci.2005.03.034
11. Vayssieres L (2003) Growth of arrayed nanorods and nanowires of ZnO from aqueous solutions. Adv Mater 15:464–466
12. Vergés MA, Mifsud A, Serna CJ (1990) Formation of rod-like zinc oxide microcrystals in homogeneous solutions. J Chem Soc Faraday Trans 86:959–963
13. Law M, Greene LE, Johnson JC, Saykally R, Yang PD (2005) Nanowire dye-sensitized solar cells. Nat Mater 4:455–459. doi:10.1038/nmat1387
14. Yang B, Lee C, Ho GW, Ong WL, Liu J, Yang C (2012) Modeling and experimental study of a low-frequency-vibration-based power generator using ZnO nanowire arrays. J Microelectromech Syst 21:776–778. doi:10.1109/JMEMS.2012.2190716
15. Tian J-H, Hu J, Li S-S, Zhang F, Liu J, Shi J, Li X, Tian Z-Q, Chen Y (2011) Improved seedless hydrothermal synthesis of dense and ultralong ZnO nanowires. Nanotechnology 22:245601
16. Woo Cho J, Seung Lee C, Il Lee K, Min Kim S, Hyun Kim S, Keun Kim Y (2012) Morphology and electrical properties of high aspect ratio ZnO nanowires grown by hydrothermal method without repeated batch process. Appl Phys Lett 101:083905. doi:10.1063/1.4748289
17. No I-J, Jeong D-Y, Lee S, Kim S-H, Cho J-W, Shin P-K (2013) Enhanced charge generation of the ZnO nanowires/PZT hetero-junction based nanogenerator. Microelectron Eng 110: 282–287

18. Govender K, Boyle DS, Kenway PB, O'Brien P (2004) Understanding the factors that govern the deposition and morphology of thin films of ZnO from aqueous solution. J Mater Chem 14:2575–2591

19. Gavrilov SA, Gromov DG, Koz'min AM, Nazarkin MY, Timoshenkov SP, Shulyat'ev AS, Kochurina ES (2013) Piezoelectric energy nanoharvester based on an array of ZnO whisker nanocrystals and a flat copper electrode. Phys Solid State 55:1476–1479. doi:10.1134/S1063783413070135

20. Tak Y, Yong K (2005) Controlled growth of well-aligned ZnO nanorod array using a novel solution method. J Phys Chem B 109:19263–19269. doi:10.1021/jp0538767

21. Greene LE, Law M, Tan DH, Montano M, Goldberger J, Somorjai G, Yang P (2005) General route to vertical ZnO nanowire arrays using textured ZnO seeds. Nano Lett 5: 1231–1236

22. Choi M-Y, Choi D, Jin M-J, Kim I, Kim S-H, Choi J-Y, Lee SY, Kim JM, Kim S-W (2009) Mechanically powered transparent flexible charge-generating nanodevices with piezoelectric ZnO nanorods. Adv Mater 21:2185–2189. doi:10.1002/adma.200803605

23. Choi D, Choi M-Y, Choi WM, Shin H-J, Park H-K, Seo J-S, Park J, Yoon S-M, Chae SJ, Lee YH, Kim S-W, Choi J-Y, Lee SY, Kim JM (2010) Fully rollable transparent nanogenerators based on graphene electrodes. Adv Mater 22:2187–2192. doi:10.1002/adma.200903815

24. Choi D, Choi M-Y, Shin H-J, Yoon S-M, Seo J-S, Choi J-Y, Lee SY, Kim JM, Kim S-W (2010) Nanoscale networked single-walled carbon-nanotube electrodes for transparent flexible nanogenerators. J Phys Chem C 114:1379–1384. doi:10.1021/jp909713c

25. Kim H, Kim SM, Son H, Kim H, Park B, Ku J, Sohn JI, Im K, Jang JE, Park J-J, Kim O, Cha S, Park YJ (2012) Enhancement of piezoelectricity via electrostatic effects on a textile platform. Energy Environ Sci 5:8932. doi:10.1039/c2ee22744d

26. Xu S, Shi Y, Kim S-G (2006) Fabrication and mechanical property of nano piezoelectric fibres. Nanotechnology 17:4497–4501. doi:10.1088/0957-4484/17/17/036

27. Chen X, Xu S, Yao N, Shi Y (2010) 1.6 V nanogenerator for mechanical energy harvesting using PZT nanofibers. Nano Lett 10:2133–2137. doi:10.1021/nl100812k

28. Wu W, Bai S, Yuan M, Qin Y, Wang ZL, Jing T (2012) Lead zirconate titanate nanowire textile nanogenerator for wearable energy-harvesting and self-powered devices. ACS Nano 6: 6231–6235. doi:10.1021/nn3016585

29. Gu L, Cui N, Cheng L, Xu Q, Bai S, Yuan M, Wu W, Liu J, Zhao Y, Ma F, Qin Y, Wang ZL (2013) Flexible fiber nanogenerator with 209 V output voltage directly powers a light-emitting diode. Nano Lett 13:91–94. doi:10.1021/nl303539c

30. Lin Y, Liu Y, Sodano HA (2009) Hydrothermal synthesis of vertically aligned lead zirconate titanate nanowire arrays. Appl Phys Lett 95:122901–122903

31. Koka A, Zhou Z, Sodano HA (2014) Vertically aligned BaTiO$_3$ nanowire arrays for energy harvesting. Energy Environ Sci 7:288

32. Xu S, Hansen BJ, Wang ZL (2010) Piezoelectric-nanowire-enabled power source for driving wireless microelectronics. Nat Commun 1:93. doi:10.1038/ncomms1098

33. Wang ZL, Song J (2006) Piezoelectric nanogenerators based on zinc oxide nanowire arrays. Science 312:242–246. doi:10.1126/science.1124005

34. Alexe M, Senz S, Schubert MA, Hesse D, Gösele U (2008) Energy harvesting using nanowires? Adv Mater 20:4021–4026. doi:10.1002/adma.200800272

35. Wang ZL (2009) Energy harvesting using piezoelectric nanowires – A correspondence on "energy harvesting using nanowires" by Alexe et al. Adv Mater 21:1311–1315. doi:10.1002/adma.200802638

36. Wang X, Song J, Liu J, Wang ZL (2007) Direct-current nanogenerator driven by ultrasonic waves. Science 316:102–105

37. Gao PX, Song J, Liu J, Wang ZL (2007) Nanowire piezoelectric nanogenerators on plastic substrates as flexible power sources for nanodevices. Adv Mater 19:67–72. doi:10.1002/adma.200601162

38. Gao Y, Wang ZL (2007) Electrostatic potential in a bent piezoelectric nanowire. The fundamental theory of nanogenerator and nanopiezotronics. Nano Lett 7:2499–2505. doi:10.1021/nl071310j
39. Gao Y, Wang ZL (2009) Equilibrium potential of free charge carriers in a bent piezoelectric semiconductive nanowire. Nano Lett 9:1103–1110. doi:10.1021/nl803547f
40. Liu J, Fei P, Song J, Wang X, Lao C, Tummala R, Wang ZL (2008) Carrier density and Schottky barrier on the performance of DC nanogenerator. Nano Lett 8:328–332. doi:10.1021/nl0728470
41. Qin Y, Wang X, Wang ZL (2008) Microfibre-nanowire hybrid structure for energy scavenging. Nature 451:809–813
42. Zhang J, Li M, Yu L, Liu L, Zhang H, Yang Z (2009) Synthesis and piezoelectric properties of well-aligned ZnO nanowire arrays via a simple solution-phase approach. Appl Phys A Mater Sci Process 97:869–876
43. Xu C, Wang X, Wang ZL (2009) Nanowire structured hybrid cell for concurrently scavenging solar and mechanical energies. J Am Chem Soc 131:5866–5872. doi:10.1021/ja810158x
44. Saravanakumar B, Mohan R, Thiyagarajan K, Kim S-J (2013) Fabrication of a ZnO nanogenerator for eco-friendly biomechanical energy harvesting. RSC Adv 3.16646. doi:10.1039/c3ra40447a
45. Xu S, Qin Y, Xu C, Wei Y, Yang R, Wang ZL (2010) Self-powered nanowire devices. Nat Nanotechnol 5:366–373
46. Zhu G, Wang AC, Liu Y, Zhou Y, Wang ZL (2012) Functional electrical stimulation by nanogenerator with 58 V output voltage. Nano Lett 12:3086–3090. doi:10.1021/nl300972f
47. Lee M, Bae J, Lee J, Lee C-S, Hong S, Wang ZL (2011) Self-powered environmental sensor system driven by nanogenerators. Energy Environ Sci 4:3359–3363
48. Yu A, Li H, Tang H, Liu T, Jiang P, Wang ZL (2011) Vertically integrated nanogenerator based on ZnO nanowire arrays. Phys Stat Solidi RRL 5:162–164. doi:10.1002/pssr.201105120
49. Hu Y, Zhang Y, Xu C, Lin L, Snyder RL, Wang ZL (2011) Self-powered system with wireless data transmission. Nano Lett 11:2572–2577. doi:10.1021/nl201505c
50. Hutson AR (1960) Piezoelectricity and conductivity in ZnO and CdS. Phys Rev Lett 4:505–507. doi:10.1103/PhysRevLett.4.505
51. Briscoe J, Stewart M, Vopson M, Cain M, Weaver PM, Dunn S (2012) Nanostructured p-n junctions for kinetic-to-electrical energy conversion. Adv Energy Mater 2:1261–1268. doi:10.1002/aenm.201200205
52. Briscoe J, Jalali N, Woolliams P, Stewart M, Weaver PM, Cain M, Dunn S (2013) Measurement techniques for piezoelectric nanogenerators. Energy Environ Sci 6:3035–3045. doi:10.1039/C3EE41889H
53. Zhu G, Yang R, Wang S, Wang ZL (2010) Flexible high-output nanogenerator based on lateral ZnO nanowire array. Nano Lett 10:3151–3155. doi:10.1021/nl101973h
54. Bai S, Zhang L, Xu Q, Zheng Y, Qin Y, Wang ZL (2013) Two dimensional woven nanogenerator. Nano Energy 2:749–753
55. Qiu Y, Zhang H, Hu L, Yang D, Wang L, Wang B, Ji J, Liu G, Liu X, Lin J, Li F, Han S (2012) Flexible piezoelectric nanogenerators based on ZnO nanorods grown on common paper substrates. Nanoscale 4:6568–6573. doi:10.1039/C2NR31031G
56. Khan A, Abbasi MA, Hussain M, Ibupoto ZH, Wissting J, Nur O, Willander M (2012) Piezoelectric nanogenerator based on zinc oxide nanorods grown on textile cotton fabric. Appl Phys Lett 101:193506. doi:10.1063/1.4766921
57. Beeby SP, Tudor MJ, White NM (2006) Energy harvesting vibration sources for microsystems applications. Meas Sci Technol 17:R175
58. Kim K-H, Kumar B, Lee KY, Park H-K, Lee J-H, Lee HH, Jun H, Lee D, Kim S-W (2013) Piezoelectric two-dimensional nanosheets/anionic layer heterojunction for efficient direct current power generation. Sci Rep 3
59. Gao Y, Zhai Q, Barrett R, Dalal NS, Kroto HW, Acquah SFA (2013) Piezoelectric enhanced cross-linked multi-walled carbon nanotube paper. Carbon 64:544–547

60. Sun H, Tian H, Yang Y, Xie D, Zhang Y-C, Liu X, Ma S, Zhao H-M, Ren T-L (2013) A novel flexible nanogenerator made of ZnO nanoparticles and multiwall carbon nanotube. Nanoscale 5:6117–6123. doi:10.1039/c3nr00866e

61. Lee M, Chen C-Y, Wang S, Cha SN, Park YJ, Kim JM, Chou L-J, Wang ZL (2012) A hybrid piezoelectric structure for wearable nanogenerators. Adv Mater 24:1759–1764. doi:10.1002/adma.201200150

62. Park K-I, Xu S, Liu Y, Hwang G-T, Kang S-JL, Wang ZL, Lee KJ (2010) Piezoelectric BaTiO₃ thin film nanogenerator on plastic substrates. Nano Lett 10:4939–4943. doi:10.1021/nl102959k

63. Seol M-L, Im H, Moon D-I, Woo J-H, Kim D, Choi S-J, Choi Y-K (2013) Design strategy for a piezoelectric nanogenerator with a well-ordered nanoshell array. ACS Nano 7:10773–10779. doi:10.1021/nn403940v

64. Kwon J, Seung W, Sharma BK, Kim S-W, Ahn J-H (2012) A high performance PZT ribbon-based nanogenerator using graphene transparent electrodes. Energy Environ Sci 5:8970–8975. doi:10.1039/c2ee22251e

65. Wu JM, Xu C, Zhang Y, Wang ZL (2012) Lead-free nanogenerator made from single ZnSnO3 microbelt. ACS Nano 6:4335–4340. doi:10.1021/nn300951d

66. Anton SR, Sodano HA (2007) A review of power harvesting using piezoelectric materials (2003–2006). Smart Mater Struct 16:R1–R21

67. Cook-Chennault KA, Thambi N, Sastry AM (2008) Powering MEMS portable devices – a review of non-regenerative and regenerative power supply systems with special emphasis on piezoelectric energy harvesting systems. Smart Mater Struct 17:43001

68. Chen X, Xu S, Yao N, Xu W, Shi Y (2009) Potential measurement from a single lead ziroconate titanate nanofiber using a nanomanipulator. Appl Phys Lett 94:253113. doi:10.1063/1.3157837

69. Cui N, Wu W, Zhao Y, Bai S, Meng L, Qin Y, Wang ZL (2012) Magnetic force driven nanogenerators as a noncontact energy harvester and sensor. Nano Lett 12:3701–3705. doi:10.1021/nl301490q

70. Batra IP, Wurfel P, Silverman BD (1973) Phase transition, stability, and depolarization field in ferroelectric thin films. Phys Rev B 8:3257–3265. doi:10.1103/PhysRevB.8.3257

71. Wurfel P, Batra IP (1973) Depolarization-field-induced instability in thin ferroelectric films - experiment and theory. Phys Rev B 8:5126–5133. doi:10.1103/PhysRevB.8.5126

72. Black CT, Farrell C, Licata TJ (1997) Suppression of ferroelectric polarization by an adjustable depolarization field. Appl Phys Lett 71:2041–2043. doi:10.1063/1.119781

73. Giocondi JL, Rohrer GS (2001) Spatial separation of photochemical oxidation and reduction reactions on the surface of ferroelectric BaTiO₃. J Phys Chem B 105:8275–8277. doi:10.1021/jp011804j

74. Kalinin SV, Bonnell DA, Alvarez T, Lei X, Hu Z, Ferris JH, Zhang Q, Dunn S (2002) Atomic polarization and local reactivity on ferroelectric surfaces: a new route toward complex nanostructures. Nano Lett 2:589–593. doi:10.1021/nl025556u

75. Dunn S, Jones PM, Gallardo DE (2007) Photochemical growth of silver nanoparticles on c- and c+ domains on lead zirconate titanate thin films. J Am Chem Soc 129:8724–8728. doi:10.1021/ja071451n

76. Dunn S, Tiwari D, Jones PM, Gallardo DE (2007) Insights into the relationship between inherent materials properties of PZT and photochemistry for the development of nanostructured silver. J Mater Chem 17:4460–4463

77. Inoue Y, Sato K, Sato K, Miyama H (1986) A device type photocatalyst using oppositely polarized ferroelectric substrates. Chem Phys Lett 129:79–81. doi:10.1016/0009-2614(86)80173-7

78. Fridkin VM (1980) Ferroelectric semiconductors. 318

79. Scott JF (2000) Ferroelectric memories. Springer, New York

80. Shao Z, Wen L, Wu D, Zhang X, Chang S, Qin S (2010) Influence of carrier concentration on piezoelectric potential in a bent ZnO nanorod. J Appl Phys 108:124312. doi:10.1063/1.3517828

81. Wang F, Seo J-H, Bayerl D, Shi J, Mi H, Ma Z, Zhao D, Shuai Y, Zhou W, Wang X (2011) An aqueous solution-based doping strategy for large-scale synthesis of Sb-doped ZnO nanowires. Nanotechnology 22:225602

82. Pham TT, Lee KY, Lee J-H, Kim K-H, Shin K-S, Gupta MK, Kumar B, Kim S-W (2013) Reliable operation of a nanogenerator under ultraviolet light via engineering piezoelectric potential. Energy Environ Sci 6:841–846

83. Keawboonchuay C, Engel TG (2003) Electrical power generation characteristics of piezoelectric generator under quasi-static and dynamic stress conditions. IEEE Trans Ultrason Ferroelectr Freq Control 50:1377–1382. doi:10.1109/TUFFC.2003.1244755

84. Sohn JI, Cha SN, Song BG, Lee S, Kim SM, Ku J, Kim HJ, Park YJ, Choi BL, Wang ZL, Kim JM, Kim K (2013) Engineering of efficiency limiting free carriers and an interfacial energy barrier for an enhancing piezoelectric generation. Energy Environ Sci 6:97–104

85. Lee S, Lee J, Ko W, Cha S, Sohn J, Kim J, Park J, Park Y, Hong J (2013) Solution-processed Ag-doped ZnO nanowires grown on flexible polyester for nanogenerator applications. Nanoscale 5:9609 9614. doi:10.1039/c3nr03402j

86. Song J, Wang X, Liu J, Liu H, Li Y, Wang ZL (2008) Piezoelectric potential output from ZnO nanowire functionalized with p-type oligomer. Nano Lett 8:203–207. doi:10.1021/nl072440v

87. Lin L, Jing Q, Zhang Y, Hu Y, Wang S, Bando Y, Han RPS, Wang ZL (2013) An elastic-spring-substrated nanogenerator as an active sensor for self-powered balance. Energy Environ Sci 6:1164–1169. doi:10.1039/C3EE00107E

88. Hu Y, Lin L, Zhang Y, Wang ZL (2012) Replacing a battery by a nanogenerator with 20 V output. Adv Mater 24:110–114. doi:10.1002/adma.201103727

89. Van den Heever TS, Perold WJ (2013) Comparing three different energy harvesting circuits for a ZnO nanowire based nanogenerator. Smart Mater Struct 22:105029. doi:10.1088/0964-1726/22/10/105029

90. Dicken J, Mitcheson PD, Stoianov I, Yeatman EM (2012) Power-extraction circuits for piezoelectric energy harvesters in miniature and low-power applications. IEEE Trans Power Electron 27:4514–4529. doi:10.1109/TPEL.2012.2192291

91. Mitcheson PD, Yeatman EM, Rao GK, Holmes AS, Green TC (2008) Energy harvesting from human and machine motion for wireless electronic devices. Proc IEEE 96:1457–1486. doi:10.1109/JPROC.2008.927494

92. Lvovich VF (2012) Impedance spectroscopy. doi:10.1002/9781118164075

93. Stewart M, Weaver PM, Cain M (2012) Charge redistribution in piezoelectric energy harvesters. Appl Phys Lett 100:73901. doi:10.1063/1.3685701

94. Wooldridge J, Blackburn JF, McCartney NL, Stewart M, Weaver P, Cain MG (2010) Small-scale piezoelectric devices: pyroelectric contributions to the piezoelectric response. J Appl Phys 107:104118. doi:10.1063/1.3380824

Index

© Joe Briscoe and Steve Dunn 2014
J. Briscoe, S. Dunn, *Nanostructured Piezoelectric Energy Harvesters*,
SpringerBriefs in Materials, DOI 10.1007/978-3-319-09632-2